FAULT LOCATION AND SERVICE RESTORATION FOR ELECTRICAL DISTRIBUTION SYSTEMS

FAULT LOCATION AND SERVICE RESTORATION FOR ELECTRICAL DISTRIBUTION SYSTEMS

Jian Liu
Shaanxi Electric Power Research Institute, China

Xinzhou Dong
Tsinghua University, China

Xingying Chen
Hohai University, China

Xiangqian Tong
Xi'an University of Technology, China

Xiaoqing Zhang
Shaanxi Electric Power Research Institute, China

Shiming Xu
NARI Technology Development Limited Company, China

This edition first published 2016

Published by John Wiley & Sons Singapore Pte. Ltd., 1 Fusionopolis Walk, #07-01 Solaris South Tower, Singapore 138628, under exclusive license granted by China Electric Power Press for all media and languages excluding Simplified and Traditional Chinese and throughout the world excluding Mainland China, and with non-exclusive license for electronic versions in Mainland China.

For details of our global editorial offices, for customer services and for information about how to apply for permission to reuse the copyright material in this book please see our website at www.wiley.com.

Library of Congress Cataloging-in-Publication Data

Names: Liu, Jian (Electrical engineer)
Title: Fault location and service restoration for electrical distribution systems / Jian Liu.
Description: Singapore : John Wiley & Sons, Inc., 2016. | Includes bibliographical references and index.
Identifiers: LCCN 2015047270 | ISBN 9781118950258 (cloth) | ISBN 9781118950272 (ePDF) |
ISBN 9781118950265 (ePUB)
Subjects: LCSH: Electric power systems–Maintenance and repair. | Electric fault location. | Electric power distribution.
Classification: LCC TK7870.2 .L58 2016 | DDC 621.319/130288–dc23 LC record available at
http://lccn.loc.gov/2015047270

Set in 10.5/13pt Times by SPi Global, Pondicherry, India
Printed and bound in Singapore by Markono Print Media Pte Ltd

1 2016

Contents

About the Authors

Professor, Ph.D., Liu Jian (b. 1967) is the Chief Engineer at the Shaanxi Electric Power Research Institute. He is also a Senior Member of IEEE. He has published over 100 papers and nine books on theory, analysis, automation and planning for distribution systems, some of which are used as text books or reference books for university students. Professor Liu's theory and algorithms have been applied in most of the Distribution Automation System (DAS) products and projects in China. As the head of the DAS Testing Group, Professor Liu is responsible for testing the fault processing performance of all the DAS built by the State Grid Corporation of China. He supervises all of the DAS standards of State Grid Corporation. Due to his excellent contribution in electric power engineering, Professor Liu has been awarded national and provincial levels of science and technology prizes 18 times.

Professor, Ph.D., Dong Xinzhou (b. 1963) is a Professor of Electrical Engineering at Tsinghua University. His research interests focus on relay protection for electric power transmission and distribution systems. He is one of the top scientists in traveling-wave based protection and its applications.

Professor, Ph.D., Chen Xingying (b. 1964) is a Professor of Electrical Engineering. She is also the Vice President of Hohai University. Her research interests focus on analysis and control of electric power distribution systems. She is one of the top scientists in self-healing control in China.

Professor Xu Shiming (b. 1967) is the Manager of the Distribution System Branch of the Nanri Electric Automation Corp. He has developed series automation products for electric power dispatching, substation automation, and distribution automation, which are widely used in China.

Professor Tong Xiangqian (b. 1961) is a Professor of Electrical Engineering at the Xi'an University of Technology. His research interests focus on analysis and automatic control of electric power distribution systems with Distributed Generations (DG).

Mr. Zhang Xiaoqing (b. 1970) is the Director of the Electric Power Research Center, Shaanxi Electric Power Research Institute. He is also a senior engineer and the assistant to Professor Liu.

Professor Shi Shenxing (b. 1975) is an Associate Professor of Electrical Engineering at Tsinghua University. He is assistant to Professor Dong.

Preface

This book is the English edition of *Fault Location and Service Restoration for Electrical Distribution Systems* published by the China Electric Power Press (Beijing, 2013), published in Chinese. It is revised and expanded and some of the contents of another book written by the authors, that is, *Relay Protection and Fault Processing for Distribution Systems* published by China Electric Power Press (Beijing, 2014), have been combined in this book, with particular emphasis on the contributions of the authors.

Fault location and service restoration for distribution systems is realized by Distribution Automation System (DAS). The authors are practitioners in China and they are experts in theory, technique, manufacturing, and engineering of DASs. Professors Liu and Zhang are from the research institute, one of the top teams on distribution automation technology and testing techniques. Professors Dong, Chen, and Tong are from universities that are famous for distribution system theory investigation. Professor Xu is from Nari Corp., which is the largest manufacturing enterprise of DASs in China. The contents of this book are the contributions of these authors.

After long-term engineering practice, the authors realized that the simpler the DAS, the more reliable it is. Thus, the authors proposed simple modeling, analysis, and optimization approaches for DASs, with which, not only is the problem of a lack of measuring data solved, but also the decision making time is reduced.

The authors also realized that the local intelligence based fault processing technologies are faster and more reliable; the distribution intelligence based fault processing technologies, such as coordination of reclosers and sectionalizers, are simple and cost effective; the centralized intelligence based DAS may locate the fault over a smaller area and restore the service in an optimized strategy, thus a complicated fault may be processed satisfactorily. Therefore, fault processing performance may be improved by coordination of the three types of fault processing technologies. The authors' experiences are the main contents of this book, which have been applied in DAS projects in over 70 cities in China.

The contents of this book are organized into six chapters. In Chapter 1, the progresses in fault processing for distribution grids are overviewed and the prospects are given. In Chapters 2–4, the interphase short circuit fault processing technologies based on local, distributed, and centralized intelligence are described, respectively. In Chapter 5, the single phase to ground fault processing technologies for neutral non effective grounding distribution systems are described. In Chapter 6, some practical aspects of fault processing are discussed, such as coordination of various fault processing approaches, planning of terminal units in DAS, and the testing and verification of fault processing performance.

In this book, Professor Liu translated and re-organized the contents of Chapters 1, 4, and 6. Professor Dong and Dr. Shi translated the contents of Chapter 5. Professor Tong translated and re-organized the contents of Chapter 2. Professor Chen translated and re-organized the contents of Chapter 3. The other authors of the Chinese version of this book carefully checked the translated manuscript.

The authors wish to thank many good friends and colleagues, including especially Professor Shen Bingbing, Professor Zhao Jianghe, Professor Liu Dong, Professor Song Guobing, Mr. Chen Yikai, Mr. Zhao Shuren, Mr. Zhang Zhihua, and Mr. Liu Bin for their encouragement and willing support, without which this book would never have been completed.

Liu Jian
Professor, Ph.D.

1

Progresses and Prospects for Fault Processing in Distribution Grids

Liu Jian

Abstract

Progresses in fault processing technologies for electrical power distribution grids are overviewed, including progresses in local, distributed, and centralized intelligence-based interphase short circuit fault location and isolation, and service restoration, as well as progresses in single-phase-to-ground fault processing. The prospects for fault processing technologies in electrical power distribution grids are discussed.

Keywords

distribution grids, overview, prospects, interphase short circuit fault, single-phase-to-ground fault, fault location, fault isolation, service restoration, relay protection, distribution automation system (DAS), feeder automation (FA)

1.1 Introduction

According to statistics, failures in distribution grids cause more than 85% of outages due to faults. Thus, fault processing technologies for distribution grids are of great importance in improving service reliability.

Faults can be divided into two categories: interphase short circuit faults and single-phase grounding faults. These faults can then be further divided into permanent and temporary faults.

Fault Location and Service Restoration for Electrical Distribution Systems, First Edition. Jian Liu, Xinzhou Dong, Xingying Chen, Xiangqian Tong, Xiaoqing Zhang and Shiming Xu.

As for earth-neutral systems, fault processing technologies for interphase short circuit faults and single-phase grounding faults are the same. However, for neutral ineffective grounding systems, such as those in China, systems are allowed to operate under single-phase grounding fault conditions for no more than 2 hours in order to ensure service reliability. The position of a single-phase grounding fault should be located and repaired in time to avoid causing an interphase short circuit fault. Interphase short circuit faults should be cleared immediately and as many affected healthy regions should be restored as quickly as possible.

The fault processing technologies can be classified into three types: (1) fault processing based on local intelligence, (2) fault processing based on distributed intelligence, and (3) fault processing based on centralized intelligence.

Fault processing approaches based on local intelligence were the earliest technologies in which neither a communication system nor master station is needed. The decision is made based solely on the information collected at the local position. Fault processing approaches based on local intelligence are still used today and include relay protection, automatic reclosing control, and backup automatic switching control. They have the advantage of fast speeds. However, the coordination of over-current protection is rather difficult in some cases, such as the feeder trunk in an urban area. Automatic reclosing control is suitable for feeders with overhead lines. Backup automatic switching control may switch the load to the backup power supplying route in several seconds, but it is only effective for loads with more than one power supplying route.

Feeder Automation (FA) based on recloser and voltage-delay type sectionalizers, reclosing with a fast over-current protection mode, and the fast healing approach based on neighbor communication, are three typical technologies of fault processing approaches based on distribution intelligence. FA based on recloser and voltage-delay type sectionalizers was invented by Japanese engineers in the 1970s and has been successfully used in Asia for several decades, but it needs reclosing twice. Reclosing with the fast over-current protection mode is an improved approach that only needs reclosing once, but requires circuit breakers instead of the former's load switches. Both FA based on recloser and voltage-delay type sectionalizers and reclosing with a fast over-current protection mode do not require communication systems and the whole feeder must undergo a period of outage. With the fast healing approach based on neighbor communication, the fault area can be located and isolated immediately and the healthy areas are hardly affected by the fault. However, high speed communication and reliability are both needed. Besides, the sectionalizing switches should be circuit breakers.

The typical technology of fault processing based on centralized intelligence is the Distribution Automation System (DAS), which consists of a master station, some sub-working-stations, a large number of Feeder Terminal units (FTU), and the communication system. Since global information can be collected, the fault location area of DAS can be much smaller and the service restoration schemes may be optimized. But DAS based fault processing needs a rather long time period, typically several minutes.

With the increasing of the amount of Distribution Generations (DG) in distribution grids, fault processing technologies coping with such challenges have been achieved.

In this chapter, the progress in fault processing technologies will be overviewed, and included is most of the literature written by the authors, which is also included in the following chapters of this book.

1.2 Progresses in Local Intelligence-Based Fault Processing

Although relay protection technologies have been used in electrical power systems for a long time, the coordination of relay protection is rather difficult in some distribution grids, such as short length urban feeders.

In many utilities, one over-current relay protection is coordinated with one or two fuses. Even on the output circuit breaker of a feeder in the substation only one over-current protection is installed. Coordination and setting of three-section overcurrent protection is investigated in References [1]–[4]. It is pointed out in [5] that interphase short circuit currents along the sectionalizing switches of a short length urban feeder are almost the same, thus the coordination of three-section overcurrent protection is difficult. An approach of time-delay coordination of the over-current relay protection scheme is suggested, in which outage on the trunk can be avoided in case of branch fails and outage on the branch can be avoided in the case of lateral fails. Four modes of hybrid schemes of three section overcurrent protection and time-delay over-current coordination are proposed in [4], which are commonly used in Chinese utilities. The coordination of over-current protection with FA based on recloser and voltage-delay type sectionalizers is described in [6].

Automatic reclosing control and backup automatic switching control also have a rather long history of application. Reference [7] describes a scheme suitable for switches on the branches or laterals of a feeder. Reference [8] describes a coordination scheme of backup automatic switching control with DAS for an area requiring high service reliability.

The local intelligence-based fault processing technology will be detailed in Chapter 2.

1.3 Progresses in Distributed Intelligence-Based Fault Processing

A family of switches with distributed intelligence are described in Reference [9] including FA based on recloser and voltage-delay type sectionalizers, FA based on coordination of reclosers, and FA based on recloser and over-current counting type sectionalizers.

FA based on recloser and voltage-delay type sectionalizers invented by the Toshiba Co. is the most widely used technology. Hai and Chen imported the technology from Japan to China and set up production lines for mass manufacture. The basic principle of FA based on recloser and voltage-delay type sectionalizers is described in References [10]–[12]. The appropriate setting of the recloser and voltage-delay type sectionalizers is the critical application problem, which is investigated based on a hierarchical model in Reference [13] and a program is also used to calculate the setting values for arbitrary grid topologies is developed.

Reclosing with a fast over-current protection mode is another distributed intelligence-based fault processing technology, the basic principle of which is described in Reference [14]. But the method in [14] has some limitations, such as long restoration time for temporary faults and enlargement of fault isolation area due to overload. Improvements are made in [15]. The duration time of temporary fault restoration is considerably reduced by adding a time delay mechanism to the tripping procedure of sectionalizers in the case of out-of-voltage. The drawback of enlarging the outage area due to overload is avoided by introducing an out-of-voltage lock mechanism into sectionalizers and loop switches, respectively. A linear planning approach is also proposed for optimizing the setting values in [15].

The approach of FA based on recloser and voltage-current mode switches is described in [16], which can be regarded as the combination of FA based on recloser and voltage-delay type sectionalizers and reclosing with a fast over-current protection mode.

These distributed intelligence-based fault processing approaches do not need communication systems and have played a great role, but they have some drawbacks, such as setting values should be adjusted in the field when the operation mode is changed.

Some distributed intelligence-based fault processing approaches with communication systems are published in [17]–[20]. A fast healing approach based on communication with GOOSE among the adjacent FTUs is described in [17], which is the typical scheme for distributed intelligence-based fault processing approaches with communication systems. The basic approach in [17] is improved in [18], in which both temporary fault and permanent fault can be located and isolated immediately without override tripping and it works well even in cases where a few switches fail to control. Other progresses also requiring communication systems are described in [19]–[20].

The distributed intelligence-based fault processing technology will be detailed in Chapter 3.

1.4 Progresses in Centralized Intelligence-Based Fault Processing

Centralized intelligence-based fault processing is the core technology of centralized intelligence-based distribution automation systems, which is always a hot topic of research, and there have been many achievements.

1.4.1 Fault Location

A unified matrix based algorithm for fault section detection and isolation in distribution systems is put forward in [21], which is improved in [22]. But the matrix based methods require both space and long calculation times for large scale distribution grids.

A fault location approach based on a directed graph is proposed in [23] without calculating a matrix. In [24], a large scale distribution grid is divided into many small scale connected systems consisting of some connected feeders and a fault may be processed in its corresponding connected system, thus the space and calculation time can be greatly reduced no matter how large scale the distribution grid is.

In [25] a hierarchical model based algorithm of fault section diagnosis for distribution networks is suggested. A fault location method based on pattern identification is described in [26]. A multi-objective distribution network restoration using an heuristic approach and a mixed integer programming method is proposed in [27]. A multi-agent based fault processing approach is described in [28].

In the field of robust fault location in case of insufficient information, there have been many achievements, which make the fault location program more usable in practice. In [29] and [30] a genetic algorithm is introduced into fault location to improve its robustness. A data mining approach based on the combination of a rough set with a neural network is described in [31] to solve the fault section identification problem in cases of insufficient information. A fuzzy reasoning approach for robust fault location in a distribution automation system is introduced in [32] and [33]. Uncertainty reasoning approaches based on Bayes probability theory for fault location in distribution grids are put forward in [34]–[37]. An integrated intelligent service restoration system for a distribution network with an auto-learning fuzzy expert system is described in [38].

As for the field of fault location for distribution systems with Distributed Generation (DG), many progresses have been reported. In [39] and [40], the influence of Distributed Generation (DG) on relay protection is investigated. The influence of DGs on types of synchro generator, asynchronous generator, and invertor on the short circuit current of distribution grids and the corresponding analyses are investigated in [41]–[43]. In [44], the suitable range of the traditional fault location approach based on over-current information for distribution systems with DGs is investigated, showing that the suitable range is rather wide, especially for overhead lines. In [45], an improved fault allocation process is proposed for overhead line-based feeders, in which the reclosing procedure and escaping DGs in fault situations are coordinated.

1.4.2 Fault Isolation and Service Restoration

Heuristic approaches are widely used to solve service restoration problems very quickly. In [46], a multi-objective distribution network restoration approach using an heuristic approach and mixed integer programming method is put forward. In [47], an optimal restoration algorithm of distribution systems using dynamic programming is described. In [48], the priority of customers is considered in service restoration. In [49], service restoration is improved through load curtailment of in-service customers. An approach to location and restoration of a short circuit with two phases grounded to the earth in non-effectively earthed distribution systems is described in [50].

Modern optimization methods [51]–[58], such as genetic algorithms, evolutionary algorithms, NSGA-II, expert-systems, Tabu searches, and so on, have been introduced into the service restoration field, which improve the restoration performance, but calculations are greatly increased.

To improve power supplying capacity, modeled topologies are built in many countries, such as three-sectioned and three-linked grids, three-supplying and one back-up grids and a 4 × 6 connection grids. But the advantages of modeled topologies may be realized by the corresponding service restoration schemes, which are described in [59] and [60].

Service restoration to avoid a breakdown over a large area – in cases of one or more bus loss of voltage due to a fault on the bus, substation, transmission line, or a transmission tower collapse – is also investigated. In [61], a mathematical algorithm for service restoration to avoid a large area breakdown is put forward. In [62] and [63], a Tabu search-based algorithm of restoration for large area blackout is described, which may form more complicated index and constraint conditions than the approach in [61]. A switching operation sequence management method is also suggested in [63]. A CSP-based model and algorithm of service restoration for large area distribution system blackout is introduced in [64]. An Agent-Environment-Rules (AER) model-based algorithm of service restoration for large-area distribution system blackout is described in [65].

Centralized intelligence-based fault processing technology will be detailed in Chapter 4.

1.5 Progresses in Single-Phase Grounding Fault Processing

A single-phase-to-ground fault is a kind of fault commonly happens in a power distribution system. Depending on the different types of neutral grounding in a power distribution system, the fault characteristics and effects are different, which require different solutions.

For single-phase-to-ground faults in power distribution systems with effective neutral grounding, zero-sequence over-current protection can detect the fault correctly. However, if it is a high impedance fault, which occurs frequently in power distribution systems, the fault characteristics are not obvious and over-current protection may malfunction. References [66]–[68] discuss high impedance faults in power distribution systems with effective earthing, zero-sequence inverse time over-current protection and third harmonic current based protection, which lays the foundation for high impedance fault detection.

A single-phase-to-ground fault generated zero sequence current is small in non-earthed power distribution system, which requires a protection-issued alarm signal, sometimes the trip signal. In order to compensate for the capacitive current when a single-phase-to-ground fault occurs, the Peterson coil is installed at earth, which may minimize the loss but makes it more difficult to protect the single-phase-to-ground faults in such a system.

For a single-phase-to-ground fault in power distribution systems with ineffectual earthing, reference [69] studies a single-phase-to-ground fault feeder selection using a zero-sequence current in all feeders radiating from the same bus bar and proposes a feeder selection method based on all feeders' current amplitudes and phase comparison. References [70]–[72] analyze single-phase-to-ground fault generated traveling waves in a fault superimposed network and verify that the initial traveling waves generated by single-phase-to-ground faults are independent of the method of neutral grounding, which presents a novel approach to solving the problem of single-phase-to-ground fault feeder selection in power distribution systems with poor earthing and a novel technology based on traveling waves.

Up to now, current traveling wave analysis based on single-phase-to-ground fault feeder selection in non-earthed power distribution systems have been widely used in the field. Traveling wave information during a permanent fault is recorded, which lays the foundations of research of fault prevention. References [73] and [74] propose the idea of single-phase-to-ground fault prevention based on traveling wave analysis. The idea can prevent faults based on fault precursors.

After a single-phase-to-ground fault occurs, accurate fault location can help maintenance staff to arrive at the fault point to remove the fault and recover power supply, which can improve power supply reliability but also reduce the workload for the line patrol and improve efficiency. References [75] and [76] analyze single-phase-to-ground fault generated initial traveling waves and present single-phase-to-ground fault location based on the time difference between single-phase-to-ground faults generated at the initial line module traveling wave and initial zero module traveling waves, which makes it possible to locate single-phase-to-ground faults in non-earthed power distribution systems.

Chapter 5 introduces single-phase-to-ground fault types and their protection strategies, focuses on high impendence fault in power distribution system with earthing through resistance and single-phase-to-ground faults in ineffective earthed power distribution systems, analyzes high impendence fault detection methods, and presents single-phase-to-ground fault feeder selection, and single-phase-to-ground fault protection, prevention, and location.

1.6 Prospects

Centralized intelligence, distributed intelligence, and local intelligence-based fault processing approaches have their respective advantages and limitations. Fault location and restoration approaches based on the coordination of centralized, distributed, and local intelligence are promising, in which the performance of fault isolation and restoration for distributed grids can be greatly improved. The coordination of centralized, distributed, and local will be detailed in Section 6.2.

The simpler a system, the more reliable it is. From another point of view, the fewer the terminal units, the more economical the system is. The planning approach to determine the amount of various kinds of terminal units to meet the reliability of service requirement is important, which will be detailed in Section 6.3 (Chapter 6).

Verification of fault processing performance of the centralized intelligence-based DAS, local intelligence-based relay protections, and distributed intelligence-based DAS is significant in guaranteeing the construction quality of DAS. The coal technology is the test technique, which will be detailed in Section 6.4.

2

Fault Processing Based on Local Intelligence

Tong Xiangqian and Liu Jian

Abstract
The connection of distributed generation in traditional single-ended source systems makes it possible to mesh the grid to a multi-ended source network, and the original three-section over-current protection with auto-reclosing will be no longer suitable. Various improved solutions to deal with the influence of grid-connected DG and inrush current from unloaded transformers, such as adding power direction elements and adopting multistage protection coordination, are proposed in this chapter.

Keywords
relay protection, reclosing, three-section over-current protection, multistage protection coordination, distributed generation, differential protection, magnetizing inrush current, second harmonic braking, coordination mode, setting principle

2.1 Introduction

This chapter discusses fault processing based on devices with local intelligence, including two types of automatic switch coordination methodologies: one without communication and one approach requiring high speed, reliable communication systems.

Fault Location and Service Restoration for Electrical Distribution Systems, First Edition. Jian Liu, Xinzhou Dong, Xingying Chen, Xiangqian Tong, Xiaoqing Zhang and Shiming Xu.

2.2 Fault Processing Based on Local Intelligence for Distribution Networks

2.2.1 Auto-Reclosure Control

The Auto-Reclosure Device (ARD) is such an autonomous device that it automatically interrupts and recloses the circuit breaker with a preset sequence of opening and reclosure when a fault occurs in the power line. The installation of an ARD in the distribution network can greatly improve the power supply reliability and reduce outage cost.

More than 80% of faults on overhead lines are temporary faults, which are caused by insulator surface flashover due to lightning, line-to-branch discharge, and/or one line touching another due to high winds or birds. When the fault line is disconnected, the dielectric strength of the fault point will be restored and the fault will be cleared automatically. Power service will be restored if the breaker recloses automatically at this moment.

ARDs can be divided into three-phase single-shot reclosing, double-shot reclosing, and tri-shot reclosing according to the reclosing times for a fault. Statistics show that up to about 80% of faults can be successfully restored by three-phase single-shot reclosing so, in China, power lines of 35 kV and less are usually equipped with three-phase single-shot reclosing devices.

ARDs should follow the following basic principles:

1. ARDs should be put into use under normal conditions and should reclose the breaker when it trips due to the relay protection device.
2. ARDs should not reclose the breaker when the operator disconnects the breaker by the control switch or the remote control device and the operator switch onto the fault manually and then trips the breaker by relay protection device.
3. The start condition of an ARD is on the principle of discrepancy, that is, the ARD gives a reclosing operation only when the control switch is on but the breaker is actually off.
4. The operating times of ARDs should be consistent with pre-specified times (such as the single-shot reclosing device can reclose the breaker only once).
5. The operating time of ARDs should be adjustable. These times should be greater than the time needed for a fault breaker and recovery of the dielectric strength of the surrounding medium, as well as for restitution and preparation for reclosing the circuit breaker and operating mechanism once more. Usually, this time is set to 0.5~1.5 s.
6. Once the ARD recloses the breaker, it should be reset automatically and ready for the next reclose.
7. The ARD should coordinate with protection devices in the system to realize the instant accelerated protection trip or the delayed accelerated protection trip.

The operating process of an ARD is as follows: for a radial distribution network, when a fault occurs, the line breaker should be opened first by protection devices, then

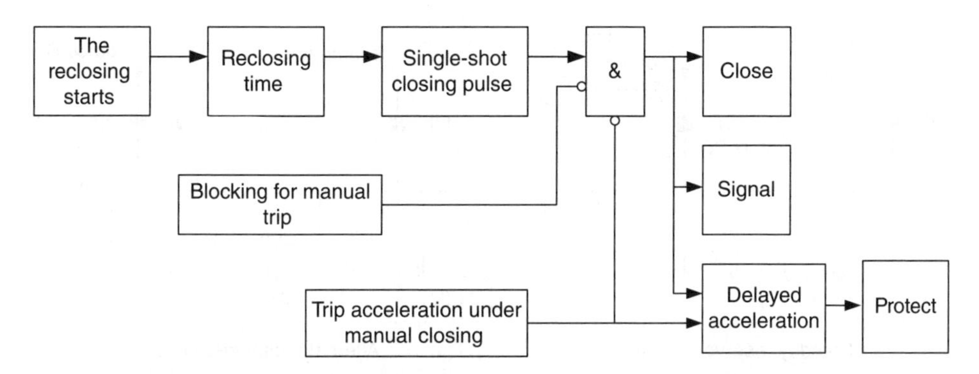

Figure 2.1 The principle diagram of three-phase single-shot reclosing

the Auto-Reclosure Device is started and recloses the circuit breaker again after a preset delay time. If the fault is temporary, reclosing will be successful and hence the power line service is restored. But if the fault is permanent, the protection system will soon let the breaker trip again after reclosure and thereafter the ARD will not reclose the breaker again.

The three-phase single-shot reclosing device is mainly composed of the starting unit, reclosing time unit, single-shot closing pulse unit, blocking device for manual trip, and trip acceleration unit under manual closure on a fault, as shown in Fig. 2.1. For a permanent fault caused by un-removed ground wire for security and unrepaired defect in the network, the re-trip of the breaker should be accelerated as much as possible to guarantee selectivity, which needs the coordinated operation of protection system and ARD.

Unlike the ARD used in the radial distribution feeder, for the ARD used in the two-source distribution network, the following two issues must be considered. (1) The ARD should reclose the breaker only after breakers on both sides of the fault point are tripped in order to extinguish the electric arc and restore the insulation strength. (2) The breaker must be closed under synchronous conditions in order to avoid electric shock at the moment of closing.

2.2.2 Automatic Backup Switching Control of the Reserve Source

The Automatic Backup Switching Control is also called as the Auto-Put-into Control. An Auto-Put-into Device of a reserve source (APD) is an automatic control device that, in the two-source system, when one source loses its voltage for some reason, it can switch the loads to another reserve-source automatically, quickly, and accurately in order to avoid interruption of the power supply and significantly improve the reliability of the electric grid.

Generally the connection of the reserve-source can be divided into standby wirings and alternate wirings, which influences the configuration of the APD, as shown in Fig. 2.2.

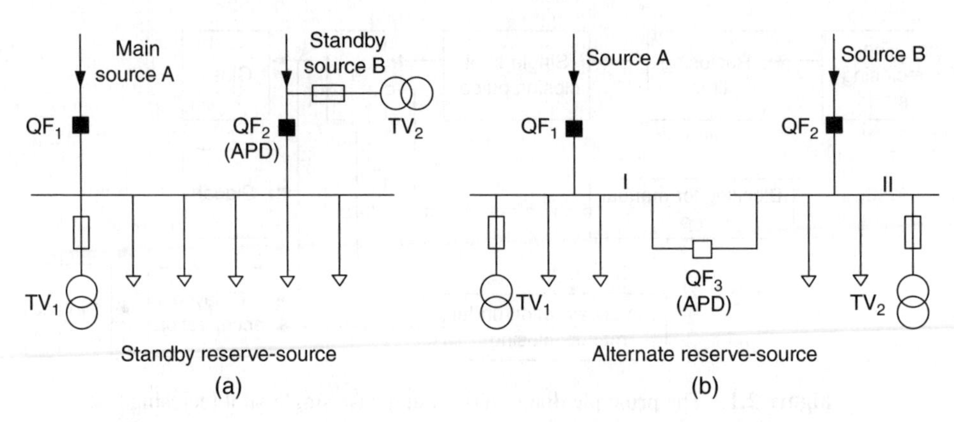

Figure 2.2 The wiring of reserve-source (a) and APD configuration (b)

There are two sources in the standby wiring; one is a main source and the other is a standby source. When the main source outage is caused by fault or maintenance, the standby source is then put to work. As shown in Fig. 2.2 (a), the APD is installed at the incoming circuit breaker QF2 of the standby source. Under normal circumstances, the bus is supplied by the main source, while the standby source is in standby state due to the breaker opening at QF2. When there is a fault in the main source, the APD will take action, the circuit breaker QF1 will be tripped, the circuit breaker QF2 will be automatically closed, and the standby source will be put to work.

There are also two sources in the alternate wirings, but they respectively provide power to their own loads under normal circumstances and mutually back each other up. When one source is out, its original load will be transferred to the other source. As shown in Fig. 2.2(b), the APD is installed at the bus-tie circuit breaker QF3. Under normal circumstances, the bus-tie breaker is in the open state, two sources supply power to the loads connecting to their own bus separately, and they are the reserve-source for each other through the breaker QF3. If bus I loses voltage due to a fault at source A, then the APD will take action, breaker QF1 will be tripped, and breaker QF3 will close automatically, thereafter the loads on bus I will be supplied by source B.

The APD of the reserve-source should comply with the following principles:

1. When one source loses its voltage, the APD should disconnect this source and then put the reserve source into service to ensure the power supply is uninterrupted.
2. The APD should not operate if a fault occurs on the load side and causes the source incoming breaker to be tripped by the protection device. The APD should not operate if the reserve source is not in service either.
3. When the working source has a planned outage, the APD shouldn't operate in order to prevent the reserve source from being put to work.
4. When the fuse of the voltage transformer is blown or the voltage transformer is disconnected by a switch, the APD should not operate.

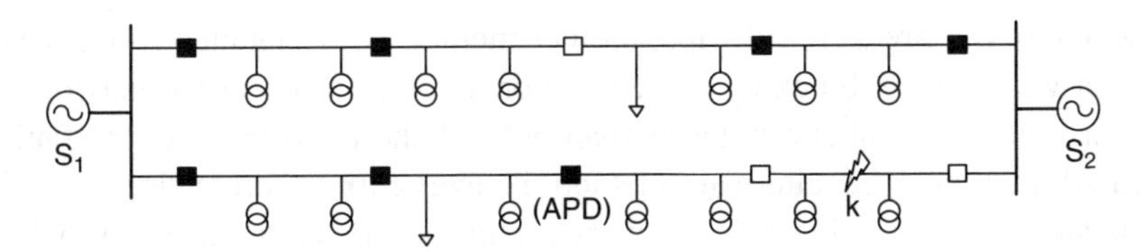

Figure 2.3 Tie switches with APD

5. The APD should operate only once, in order to avoid the reverse source switching on a permanent fault.
6. The operating time of APD should be as short as possible.

For the tie switch or the section switch in a meshed distribution network, the APD should be allocated. When the original source has an outage because of a source failure, the tie switch or section switch will be turned on automatically by the APD, thereafter the power service in all areas will be restored automatically by another source on the other side.

For example, as shown in Fig. 2.3, when the point k of the power line has a permanent fault, the fault area will be isolated by the switch on both sides, then the tie switch with the APD will be closed automatically, and restore power service for the area without fault.

2.2.3 Voltage Protection

Voltage protection includes over-voltage protection and under-voltage protection.

2.2.3.1 Over-Voltage Protection

There are two kinds of over-voltage in the distribution network according to the causes of lightning or switch operation. Over-voltage will occur when a lightning strike hits a power line or through a transient process caused by a switch operation in the distribution system.

The means of over-voltage protection in the power line is usually by installing a lightning arrester on the distribution bus or the area close to the protected region. For some voltage-sensitive devices (e.g., electronic devices, distributed generation, etc.), over-voltage protection relays can be installed. When the network voltage is larger than the threshold value, those voltage-sensitive devices will be split from the grid.

2.2.3.2 Under-Voltage Protection

Under-voltage in the distribution network is usually caused by grid fault or overload. One method of under-voltage protection is load shedding in the under-voltage condition. As long as the bus voltage is lower than the setting value (e.g., 60% of the rated voltage), the breaker will trip. Another method is instantaneous under-voltage protection with current blocking. When a short circuit occurs in the network, if the important users' bus voltage is lower than 60% of the rated voltage, the fault must be cleared rapidly. The alternate method is over-current protection coordinated with under-voltage starting, if an over-current occurs simultaneously when the voltage decreases or the negative sequence voltage increases, the fault must be cleared rapidly.

For example, provided that the definite time over-current protection and the instantaneous over-current protection are installed at a feeder, because the setting value of instantaneous over-current protection is larger than the maximum short circuit current at the end of the power line, there is a rather large dead zone of the instantaneous protection and hence low sensitivity. When a fault occurs at the rear end of the feeder, it can only be cleared by definite time over-current protection. Under this condition, in order to improve sensitivity and reduce the dead zone of protection, the setting current can be reduced but, meanwhile, an additional under-voltage start-up condition must be adopted (e.g., the voltage is lower than the 60% of the rated voltage) to insure the reliability of instantaneous over-current protection.

2.2.4 *Three-Section Over-Current Protection*

Current protection is the most important ways to protect the distribution network. Almost all faults are accompanied with a surge of current in the power line. At present, there are two main schemes of current protection in the distribution network, three-section over-current protection and inverse-time over-current protection.

Three-section over-current protection is composed of instantaneous over-current protection (Section I), time delay instantaneous over-current protection (Section II), and definite time over-current protection (Section III). Figure 2.4 shows the operation logic of three-section current protection, where KA denotes the current relay and KT denotes the time relay.

Instantaneous over-current protection is generally set to a value that is larger than the maximum short-circuit current when the fault occurs at the end of the power line. The instantaneous protection will clear the fault quickly, but it can't protect the whole length of the line and will leave a dead zone of protection.

Time delay instantaneous over-current protection should protect the whole length of the line segment and therefore be set according to the principle that there is enough sensitivity when a fault occurs at the end of its protection zone and this protection should also coordinate with the instantaneous current protection of downstream zone of the line. However, this protection has a time limit to meet the requirement of protection selectivity.

Definite-time over-current protection should also protect the whole length of the line segment and therefore the setting value should be greater than the maximum

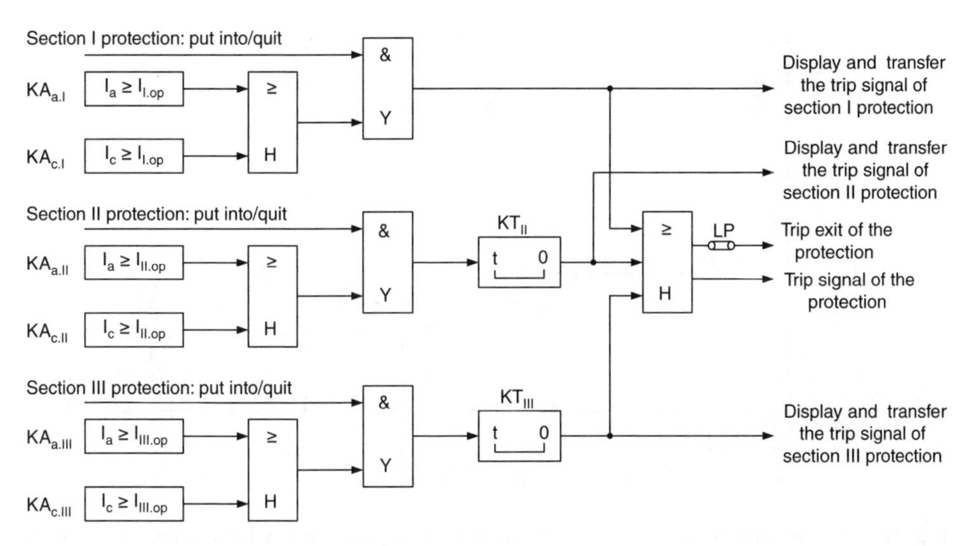

Figure 2.4 The operation logic of three-section current protection

load current in this line. Definite-time over-current protection for this line segment should also coordinate with the definite-time over-current protection for the downstream segment by an operating time delay to meet the protection selectivity.

Inverse-time over-current protection protects by making the trip time inversely proportional to the short-circuit current in the protected line. That is, the shorter the distance between the fault point and the protection position, the larger the short-circuit current, and the longer the time limit of the protection. The inverse-time over-current protection can satisfy the requirements of both rapidity and selectivity.

Definite-time over-current protection and inverse-time over-current protection have their own characteristics. The hardware system of the definite-time over-current protection is more complex than the inverse-time, but the coordination of definite-time over-current protection between the upstream and downstream protection is much easier than inverse-time. With the popularity of computer protection, hardware differences between the protection devices no longer exist, so definite-time over-current protection is more commonly used. However, Section III of the three-section current protection can usually be configured as definite-time or inverse-time over-current protection. Inverse-time protection is discussed no further in this chapter.

It is worth pointing out that the vast majority of faults in the distribution network are temporary faults. Therefore, for an overhead line network, whatever protection scheme is used, it is necessary to install an ARD to restore power service in the case of a temporary fault.

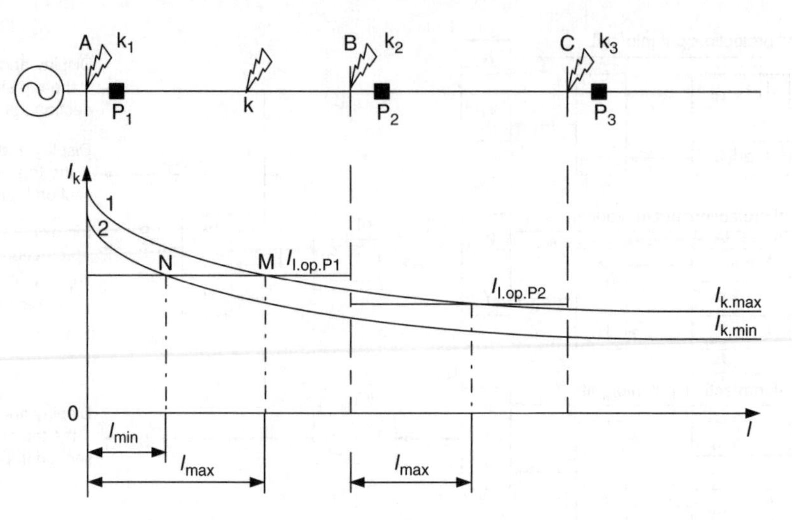

Figure 2.5 Instantaneous over-current protection of single-ended radial distribution network

2.2.4.1 Instantaneous Over-Current Protection (Section I)

Figure 2.5 shows a single-ended radial distribution network. Where curve 1 is the maximum short circuit current $I_{k.max}$ (the three-phase short-circuit current in the maximum operating mode of the grid) that varies with fault position k (apart from the source bus A), and curve 2 is the minimum short-circuit current $I_{k.min}$ (the two-phase short-circuit current in the minimum operating mode of the grid) that varies with fault position k.

Instantaneous over-current protection emphasizes selectivity and rapidity. It will disconnect the power line and clear the fault without delay only when a fault occurs on this segment of the line. When instantaneous over-current protection device P1 is installed to protect the line segment AB, P1 must meet the following requirements: P1 should not trip when a fault occurs on the outside of line segment AB (such as a fault occurs on the line segment BC), and P1 must trip when a minimum fault occurs on the head of the line segment AB. Therefore, the operating current of instantaneous over-current protection at P1 should satisfy the following relationship:

$$I_{k2.\max} < I_{I.op.P1} < I_{k1.\min}$$

Where

$I_{k1.min}$ = the minimum short-circuit current at the point k1, that is the short-circuit current when a fault occurs on the point k1 and the grid is in a minimal operation state.

$I_{k2.max}$ = the maximum short-circuit current at the point k2, that is the short-circuit current when a fault occurs on the point k2 and the grid is in a maximal operation state.

$I_{I.op.P1}$ = the current setting value of section I or instantaneous over-current protection at P1.

Generally, the operating current of instantaneous over-current protection is set to a value that is larger than the maximum three-phase short-circuit current of the fault occurring on the end of this segment, that is

$$I_{\text{I.op.P1}} = K_{\text{rel}}^{\text{I}} I_{\text{k2.max}} \tag{2.1}$$

Where, $K_{\text{rel}}^{\text{I}}$ is the reliability coefficient of instantaneous over-current protection, generally $K_{\text{rel}}^{\text{I}} = 1.2 \sim 1.3$.

The minimum two-phase short-circuit current at the head of the line is used to verify the sensitivity of this protection,

$$K_{\text{sen}} = \frac{I_{\text{k1.min}}}{I_{\text{I.op.P1}}} \tag{2.2}$$

Where, K_{sen} is the sensitivity coefficient, generally $K_{\text{sen}} \geq 2$.

The actual operating time of instantaneous over-current protection is determined by the inherent opening and operating time of the protection device, which includes the protection exit relays and circuit breakers. The inherent operating time of the exit relay is no more than 25 ms, and the inherent opening time of a circuit breaker is mostly in 30~60 ms. Let t_0 denote the inherent opening time of the protection device, which means $t_{\text{I}} = t_{0.}$

As shown in Fig. 2.5, instantaneous over-current protection cannot protect the whole segment to be protected. l_{max} and l_{min} are the protection range of instantaneous over-current protection under maximum and minimum short-circuit current, respectively. For instance, protection P1 will not trip immediately when a short circuit occurs at the end of the line AB. It means that instantaneous over-current protection has a protection dead zone, and its range depends on the difference of short-circuit current between the head and the end of the protected line. Instantaneous over-current protection may have a long dead zone and then make the protection lose its function in the condition of a short line, a small short-circuit capacity of the system, or big changes to the system operation mode. The faults in the dead zone must be accomplished by time delay instantaneous over-current protection or definite- time over-current protection.

2.2.4.2 Time Delay Instantaneous Over-Current Protection (Section II)

Time delay instantaneous over-current protection is designed to protect the whole segment of the line and to compensate for the dead zone of instantaneous over-current protection. Time delay instantaneous over-current protection can not only be used as a main protection, but also as a backup protection to instantaneous over-current protection.

Taken the network shown in Fig. 2.5 as an example, assuming that time delay instantaneous over-current protection for the line AB is installed to P1, the protection P1 should trip when the minimum short circuit occurs in the line AB and, meanwhile, the range of protection should be no more than the protective range of instantaneous over-current protection for the downstream line BC (section I in P2). Therefore, the operating current of time delay instantaneous over-current protection at P1 should satisfy the following relationship

$$I_{\text{I.op.P2}} < I_{\text{II.op.P1}} < I_{\text{k2.min}}$$

Where

$I_{\text{I.op.P2}}$ = the current setting value of instantaneous over-current protection at P2 (section I).
$I_{\text{II.op.P1}}$ = the current setting value of time delay instantaneous over-current protection at P1 (section II).

Generally, the current setting value of time delay instantaneous over-current protection must be coordinated with instantaneous over-current protection in the downstream line,

$$I_{\text{II.op.P1}} = K_{\text{rel}}^{\text{II}} I_{\text{I.op.P2}} \tag{2.3}$$

And sensitivity can be verified by

$$K_{\text{sen}} = \frac{I_{\text{k2.min}}}{I_{\text{II.op.P1}}} \tag{2.4}$$

Meanwhile, the time limit of the time delay instantaneous over-current protection must be greater than the instantaneous over-current protection in the downstream line: a time difference to meet the selectivity of protection.

$$t_{\text{II.P1}} = t_{\text{I.P2}} + \Delta t \tag{2.5}$$

Where

$K_{\text{rel}}^{\text{II}}$ = reliability coefficient of time delay instantaneous over-current protection, usually $K_{\text{rel}}^{\text{II}} = 1.1$
K_{sen} = sensitivity coefficient, generally $K_{\text{sen}} \geq 1.5$
$t_{\text{I. P2}}$ = time limit of instantaneous over-current protection at P2
$t_{\text{II.P1}}$ = time limit of time delay instantaneous over-current protection at P1
Δt = time difference of protection, generally $\Delta t = 0.3$~0.5 s

Figure 2.6 shows the coordination between time delay instantaneous over-current protection for this segment and instantaneous over-current protection for the next segment.

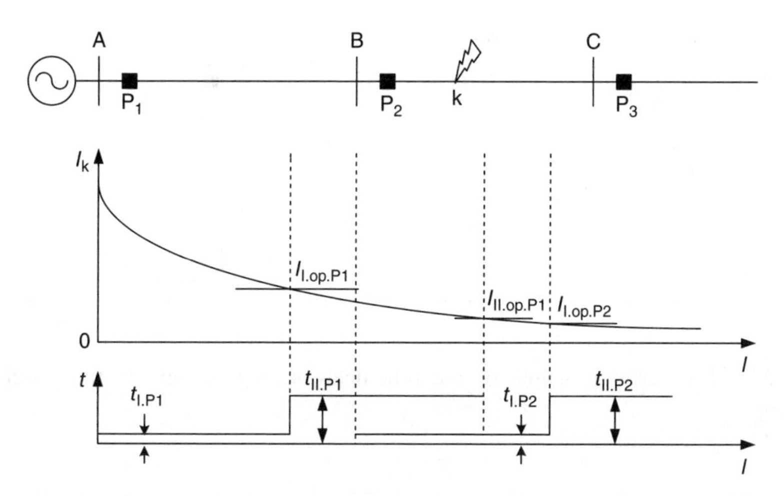

Figure 2.6 Coordination of time delay instantaneous over-current protection with instantaneous over-current protection for next segment

When a fault occurs at the point k on the line BC, the fault current will start both the instantaneous over-current protection in P2 and the time delay instantaneous over-current protection in P1. However, due to the longer time limit of time delay instantaneous over-current protection in P1, the fault will be removed selectively by the instantaneous over-current protection in P2. If P2 refuses to trip for some reason, P1 will clear the fault with a time delay and without selection.

It is worth pointing out that the time delay instantaneous over-current protection may be of low sensitivity and may not protect the whole line where the line is short, there is a small short-circuit capacity of the system, or there are big changes to the system operation mode. Therefore, the definite time over-current protection must be added as a backup protection.

2.2.4.3 Definite Time Over-Current Protection (Section III)

Definite time over-current protection is generally used as the local backup protection when main protection in this line refuses to trip and also as the remote backup protection for the next segment of the line. Definite time over-current protection must be able to protect the full length of this segment. Compared with the current protection of section I and section II, Section III has the lowest current setting and the highest sensitivity. The current setting of definite time over-current protection should be larger than the maximum load current, that is

$$I_{\mathrm{III.op.P1}} = \frac{K_{\mathrm{rel}}^{\mathrm{III}} K_{\mathrm{st}}}{K_{\mathrm{re}}} I_{\mathrm{L.max}} \tag{2.6}$$

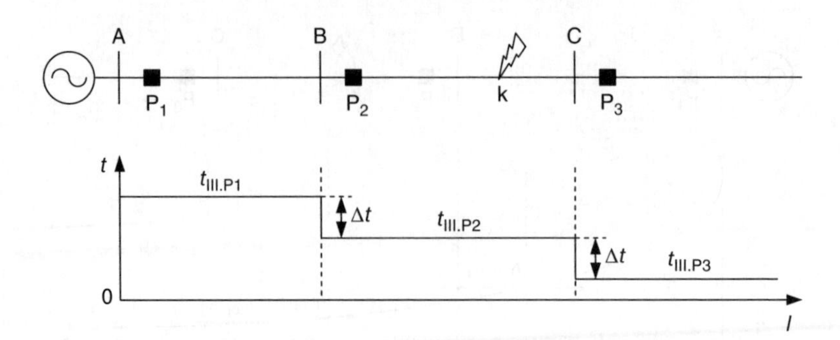

Figure 2.7 The scalar principle of the time-delay setting of definite time over-current protection

And sensitivity of the protection is verified by the two-phase short-circuit current at the line end under the minimum operating mode,

$$K_{sen} = \frac{I_{k2.min}}{I_{III.op.P1}} \tag{2.7}$$

Selectivity of definite time over-current protection is ensured through the time coordination between the two adjacent segment protections. As shown in Fig. 2.7, according to the time scalar principle, the operating time of section III is set to

$$\begin{cases} t_{III.P1} = t_{III.P2} + \Delta t \\ t_{III.P2} = t_{III.P3} + \Delta t \end{cases} \tag{2.8}$$

where

K_{rel}^{III} = reliability coefficient of definite time over-current protection, generally K_{rel}^{III} = 1.2 ~1.3
K_{st} = self-start coefficient of motors
K_{re} = return coefficient of current relay, generally K_{re} = 0.85~0.9
$I_{III.op.P1}$ = current setting of definite time over-current protection of P1 (section III)
$I_{L.max}$ = maximum load current in the protected line
K_{sen} = sensitivity coefficients, required no less than 1.5 when the fault is on the end of this segment, and no less than 1.2 when the fault is on the end of the next segment
$t_{III.P1}$ = time delay of definite time over-current protection of P1
$t_{III.P2}$ = time delay of definite time over-current protection of P2
$t_{III.P3}$ = time delay of definite time over-current protection of P3.

In order to improve the rapidity of over-current protection, the current protection of section III can also adopt an inverse-time over-current protection. According to the

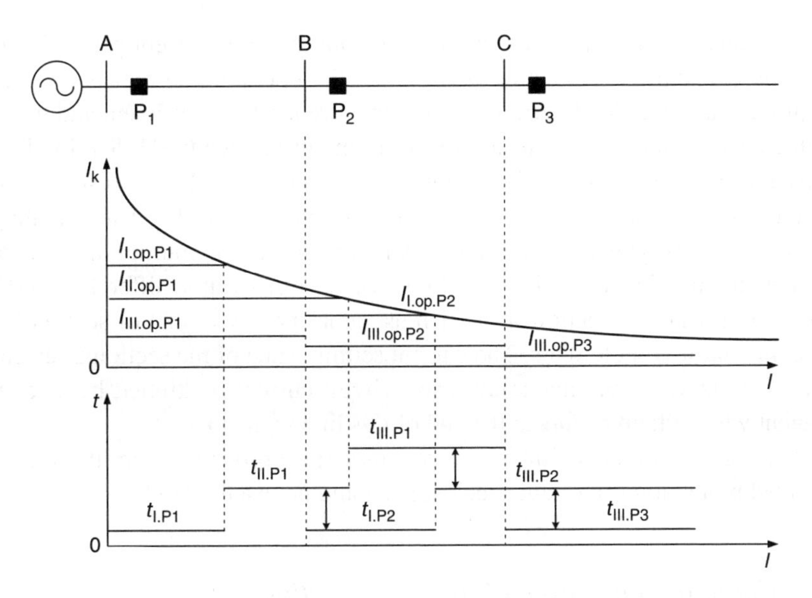

Figure 2.8 The coordination of three section over-current protection for a multi-segmented feeder

IEC255–4 from the International Electrotechnical Commission, based on the current setting and time setting of the section III current protection, the characteristic of the inverse-time over-current protection is generally expressed as

$$t = \frac{0.14}{\left(I / I_{\text{III.op}}\right)^{0.02} - 1} t_{\text{III}} \tag{2.9}$$

2.2.4.4 The Coordination of Three-Section Current Protection

For a single feeder with multi segmented breakers, current protection is installed with all three sections at every circuit breaker. Taking the three segmented feeder shown in Fig. 2.8 as an example, the protection configuration should comply with the following rules: (1) To design gradually from the terminal of the feeder to the beginning of the feeder; (2) Definite time over-current protection (section III) should be installed first, then the instantaneous over-current protection (section I), and last, time delay instantaneous over-current protection (section II); and (3) If the time limit of definite time over-current protection (section III) is no more than 0.5 s, instantaneous over-current protection will be unnecessary.

Only one section, section III, is configured to current protection of P3. The operating current is set to a value of the maximum load current multiplying a reliability coefficient and the time delay is setting to 0 s. It means that the delay time is only the inherent trip time of the protection device.

Two sections, section III and section I, are configured to current protection of P2. First section III current protection is installed, the operating current is also set to a value of the maximum load current multiplying a reliability coefficient and the operation time delay is set to 0.5 s, which forms a time difference to P3. Section I current protection is also installed in order to improve protection rapidity for a serious fault.

All three sections are configured to the current protection of P1. The operating current of section III is set to a value of the maximum load current multiplying a reliability coefficient and the time delay is set to 1.0 s, which forms a time difference to P2. The current setting value of section II of P1 must coordinate with that of section I of P2, and the time limit is set to 0.5 s. The current setting value of the section I current protection of P1 is set to the maximum short circuit current multiplied by a reliability coefficient when a fault occurs at the end of this line segment.

Furthermore, in order to improve temporary fault processing ability, the instant accelerated protection trip of auto-reclosure should be installed to P1.

2.2.5 Coordination between Current Protection Relaying and Auto-Reclosure

Statistics show that more than 80% of faults on overhead lines are temporary faults. In order to restore power service as quickly as possible after tripping due to a fault in the distribution network, auto-reclosing devices (ARDs) are usually installed on overhead lines to improve the reliability of power supply in the distribution network. In a radial line with multi-stage protection, the automatic reclosing device should be able to coordinate with protection relays to achieve an instant and delayed accelerated protection trip.

2.2.5.1 Instant Accelerated Protection Trip

In instant accelerated protection trip, the ARD is only installed to the nearest stage of protection to the power source. As shown in Fig. 2.9, provided that definite time overcurrent protection is installed at all stage protections of P1, P2, and P3, their operating delay time coordinates by the time scalar principle. No matter where a fault occurs along the line, the fault will be cleared first by the nearest P1 protection breaker without time delay or selectivity. After that, the automatic reclosing device will reclose the P1 breaker only once, the reclosing will be successful, and power service will

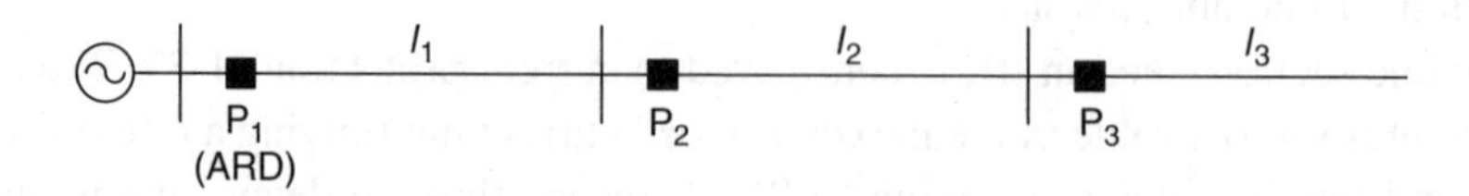

Figure 2.9 Instant accelerated protection trip

be entirely restored if the fault is temporary. But if the fault is permanent, the fault will be cleared selectively by the corresponding protection device with a time delay. Meanwhile, auto-reclosure will be blocked.

Instant accelerated protection trip requires only one ARD and has simple, economical, and rapid characteristics. It can also avoid making a temporary fault into a permanent fault. However, if the fault is permanent, the feeder and loads will suffer short-circuit impact twice over and more time is needed for the second clearing of fault. Instant accelerated protection trip is mainly used in the feeders directly from the substation up to 35kV.

2.2.5.2 Delayed Accelerated Protection Trip

In the delayed accelerated protection trip, protection devices along every segment of the line are configured with a three-phase auto-reclosing function, as shown in Fig. 2.10. When a fault occurs on a segment of the protected line, the protection device corresponding to this segment will clear the fault selectively. After then, the ARD will only reclose the breaker once. The reclosing will be successful and power service entirely restored if the fault is temporary, but if the fault is permanent, the protection device will accelerate the trip, let the breaker trip again without time delay and block this auto-reclosure at the same time.

The delayed accelerated protection trip adopts more ARDs and therefore is expensive. It also needs more time to remove the fault initially. However, the first trip is selective and so the blackout area is reduced. The delayed accelerated protection trip is usually used in the 35 kV and over line, feeding to important loads.

2.2.6 Directional Over-Current Protection

For a two-source distribution line or a single-source ring-network, both ends of the line need a circuit breaker configured to current protection along the power flow direction to meet the requirement of selectivity and isolating the fault area from other similar areas of the network.

Taking the single-source ring-network shown in Fig. 2.11 as an example, P1~P10 are the segment breakers with protective functions and tie switch LK is closed under normal conditions. Define the power direction from the source to loads as positive, then P1, P3, P5, P7, P9 will provide positive direction over-current protection for

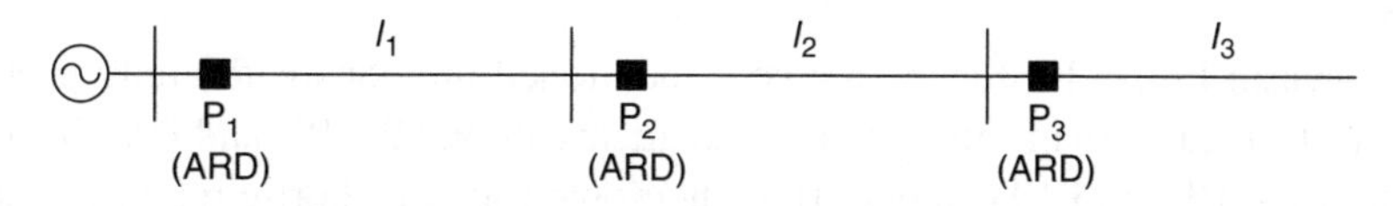

Figure 2.10 Delayed auto-reclosing protection trip

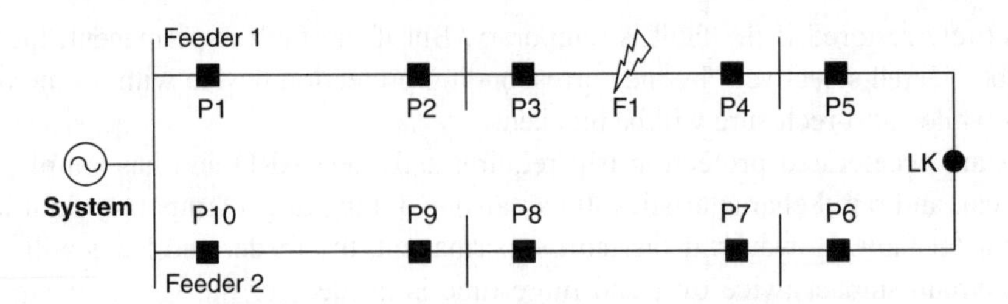

Figure 2.11 Directional over-current protection

feeder 1, and P10, P8, P6, P4, P2 will provide positive direction over-current protection for feeder 2. Only the forward short-circuit current flowing from the protection device can start protection. When a fault occurs on the line (such as F1 in Fig. 2.11), both breakers around the fault (such as P3 and P4) must be disconnected by directional over-current protection to isolate the fault from other non-faulted segments of the ring-line.

The setting principles of an operating current with directional over-current protection are the same as those for definite time over-current protection and selectivity is guaranteed by a time delay complying with the scalar principle. The time delay of the protection devices in the ring-line should be satisfied by

$$t_1 > t_3 > t_5 > t_7 > t_9$$
$$t_{10} > t_8 > t_6 > t_4 > t_2$$

If a short circuit occurs at F1, P1 and P3 will start due to the forward short-circuit current from feeder 1, but only P3 will trip because P3 has a smaller time delay than P1. On the other hand, P10, P8, P6, and P4 will start due to the forward short-circuit current from feeder 2, but only P4 will trip because P4 has the smallest time delay of the four protection devices. Finally, the fault at F1 is isolated by both P3 and P4 with good protection selectivity.

Directional over-current protection can reliably isolate the fault to the smallest area, but its shortcomings are also obvious. First, the closer the fault point is to one source, the longer the tripping time will be. Just as in the example mentioned previously, the operating time t_3 of P3 is longer than time t_4 of P4. Second, directional over-current protection needs an additional directional element and voltage transformer. In particular, the directional element may lose its function if the short circuit causes a very low voltage detected by the protection device. So it is necessary to reduce directional elements.

As shown in Fig. 2.12, for the network mentioned previously, the only protection points that truly need to install directional elements are P2, P4, P7, and P9. The reasons are as follows: P1 and P10 can only flow their own forward short-circuit current from feeder 1 and feeder 2, respectively, and the time delay is also the longest, so P1 and P10

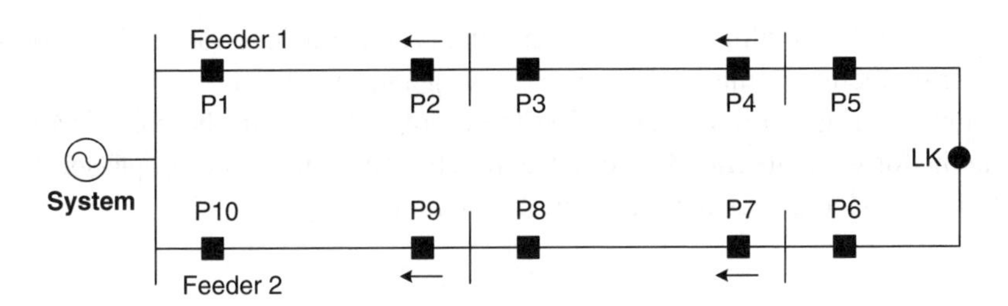

Figure 2.12 The necessity of directional element

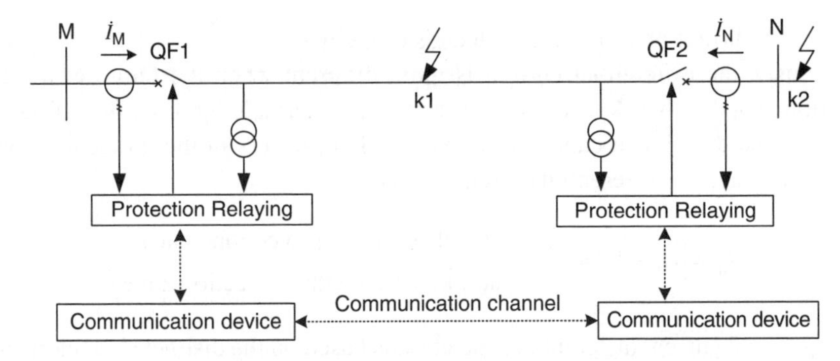

Figure 2.13 The schematic diagram for longitudinal current differential protection

are unnecessary for installing the directional element. The time delay of P3 is longer than P2. Even if a fault occurs between P1 and P2, P2 will precede P3 to take correct action. Therefore, P3 is unnecessary for installing the directional element. For the same reason, directional elements of P5, P6, and P8 are also unnecessary.

2.2.7 Longitudinal Current Differential Protection

For the over-current protection for single-ended source system mentioned previously, to ensure selectivity, instantaneous over-current protection can only protect a part of the line, so the fault in the dead zone can only be removed by protection with a time limit. But this kind of delay trip is undesirable or not permissible for the important single-ended source system and especially for the double-ended source system. Longitudinal current differential protection can instantaneously clear a fault in the whole protection zone.

By comparing the value and phase of currents flowing through the two current transformers on both sides of the line, longitudinal current differential protection can judge whether a fault occurs in the protection zone and will then make instantaneous trip protection. As shown in Fig. 2.13, longitudinal current differential protection needs to install protection devices at both ends of the protected line, and the communication channel between both ends should be established to exchange current information each other. However, longitudinal current differential protection does not need to coordinate with other protection points in the current setting and time limit.

As shown in Fig. 2.13, we can define the current flow from the bus of M to N as going in the forward direction first of all, and define the sum of current phasor detected on both bus of M and N as the differential current, denoted I_d

$$I_d = |\dot{I}_M + \dot{I}_N| \tag{2.10}$$

Where

$\dot{I}_M$, $\dot{I}_N$ = The current phasor at the both ends of the line

Obviously, the differential current is quite different when the fault occurs in the protection zone (such as k_1) or outside the zone (such as k_2). Ignoring the influence of line-to-ground distributed capacitance, when a fault occurs in the protection zone or outside the zone, the differential current will be

$$I_d = |\dot{I}_M + \dot{I}_N| = \begin{cases} I_{k1} & (\text{fault k1 in the protection zone}) \\ 0 & (\text{fault k2 outside the protection zone}) \end{cases}$$

Longitudinal current differential protection is based on the distinct characteristics of differential current.

Due to the influence of distributed capacitance of the line, the error of current transformers on both ends of the line, and differences in characteristic of the current relays, the differential current will not be zero during normal conditions, especially for an external fault. This non-zero current is called the unbalanced differential current, and the larger the external fault current is, the larger the unbalanced differential current will be.

To improve the sensitivity of the differential protection for an internal fault and also to suppress the influence of unbalanced current for an external fault, a braking current, denoted I_r, is defined as,

$$I_r = |\dot{I}_M - \dot{I}_N| \tag{2.11}$$

Therefore, the trip of differential protection should meet the following conditions simultaneously

$$\begin{cases} I_d > I_0 \\ I_d > k_r I_r \end{cases} \tag{2.12}$$

Where

k_r = ratio braking coefficient, usually $k_r = 0.6$

I_0 = the maximum unbalanced current during normal conditions, usually set to 0.4 times the calculated load current of the line.

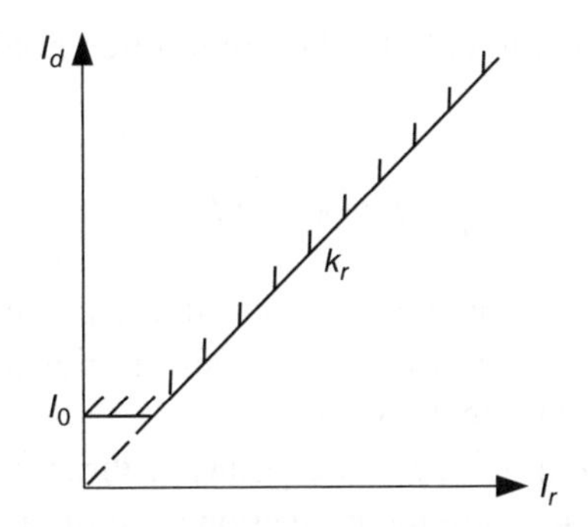

Figure 2.14 The characteristic of differential current protection with ratio restraint

Figure 2.14 shows the characteristics of differential current protection with ratio restraint, and the ratio braking coefficient is the slope of the dashed line. The differential current protection should trip when the differential current lies above the dashed line. The double-ended and single-ended source systems will be analyzed, respectively, in the following.

2.2.7.1 Double-Ended Source System

Provided that the short-circuit currents coming from two sources are the same when the short circuit occurs in the protection zone, both currents are denoted I_k. If a fault occurs in the protection zone, the short-circuit current that flows through both M and N buses will be of the same value and the same phase, so the differential current will be $2I_k$ and the braking current will be zero. Obviously, formula (2.12) is tenable and the differential current will reliably fall in the trip zone. But if a fault occurs outside the protection zone, the short-circuit current that flows through both M and N buses will be of the same value and opposite phase, hence the differential current will be zero and the braking current will be $2I_k$. In this case, (2.12) will no longer be tenable.

2.2.7.2 Single-Ended Source System

If a fault occurs in the protection zone, the fault current I_k provided by the source will flow only through the M bus, while no current will flow through the N bus. So the differential current will be I_k and the braking current will also be I_k; obviously, Equation (2.12) will be tenable. If a fault occurs outside the protection zone, fault current I_k provided by the source will flow through both M and N buses. So the differential current will be zero and the braking current will be $2I_k$; obviously, Equation (2.12) will no longer be tenable.

This analysis shows that longitudinal current differential protection has higher sensitivity in a double-ended source system than in a single-ended source system.

The starting condition of the longitudinal current differential protection shown in the formula (2.12) can be rewritten as:

$$I_d - k_r I_r \geq I_0 \tag{2.13}$$

It is worth pointing out that the unbalanced current must be greater than the maximum inrush current of transformer if power transformers are contained in the longitudinal differential protection zone.

Furthermore, the direction of the power flowing through both ends of the line is different when a fault is located in the protection zone or outside the protection zone. When a fault occurs in the protection zone, the short-circuit power of the two ends flows from both buses to the fault point, then both power flows are in the same direction. But when a fault occurs outside the protection zone, the power flows on the two buses are in opposite directions, one short-circuit flows out of its bus and the other flows into its bus. Based on this power directional characteristic, longitudinal current directional protection can also be constituted.

2.2.8 The Second Harmonic Braking Criterion in Current Protection

The magnetizing inrush current of a transformer depends on the magnetic characteristics of its core, the remnant flux, the closing initial phase angle, and so on. Under normal conditions, steady main magnetic flux is established in the core of transformer and the magnetic flux density cannot be saturated in the core, so the magnetizing current is very small, usually taking up about 2–10% of the rated current of the transformer. However, in the temporary process of restorative closing after fault removal or due to a unloaded transformer energizing, the magnetic flux density of the transformer will increase in saturation, resulting in a sharp increase in magnetizing current, the value of which may reach 4~8 times the rated current of transformer.

The magnetizing inrush current may cause incorrect protection tripping, especially for the longitudinal current differential transformer protection and the three-section current protection of a distribution line. Increasing the current protection setting value will decrease or avoid the effect of the magnetizing inrush current of transformer on protection stability, but the protection sensitivity will be reduced. The addition of a time delay on the breaker or over-current protection reclosing device can also prevent false tripping caused by an inrush current at the closing moment, but it also affects rapidity of protection. Therefore, how to identify the magnetizing inrush current from the fault current and distinguish the two is an effective method to find solutions to the influence of magnetizing inrush current on current protection.

By analyzing the magnetizing inrush current of three-phase transformer, it shows that at least one phase current of the three-phase magnetizing inrush current contains a rather large second harmonic component. Therefore, several blocking criteria for magnetizing inrush current based on the second harmonic braking principle have been adopted to prevent false tripping caused by inrush current.

Currently, the main criteria for blocking inrush currents based on the second harmonic braking principle are as follows.

2.2.8.1 Criterion 1

Take the ratio of the second harmonic component to the fundamental component of the maximum current among the three phase currents as an index. If the index is larger than its setting value, there is a magnetizing inrush current and the trip should be blocked.

$$I_2 / I_{1.\max} \geq K_{set} \tag{2.14}$$

Where

$I_{1.\max}$ = fundamental component of the maximum current among the three phase currents

I_2 = 2nd harmonic component of the maximum current among the three phase currents

K_{set} = setting value of second harmonic braking ratio, usually $K_{\text{set}} = 0.15 \sim 0.17$.

Criterion 1 uses the information from the fault phase; hence, it is often called fault phase braking. When an unloaded transformer with fault is energized, the current protection will quickly and accurately trip to protect the transformer. However, for the three-phase transformer with the Yd connection, the current measured in the Y side actually is usually the difference of two phase current to ensure the balance of differential current between the Y side and the d side. If the symmetrical inrush current occurs, the second harmonic current may be very small and result in braking failure.

2.2.8.2 Criterion 2

Calculate the ratio of the second harmonic component to the fundamental component of each phase current, and take the maximum ratio as an index. If the index is larger than its setting value, there is a magnetizing inrush current and the trip should be blocked.

$$\begin{cases} (I_2 / I_1)_{\max} \geq K_{set} \\ I_1 > I_{op.2} \end{cases} \tag{2.15}$$

Where

I_1 = fundamental current
I_2 = 2nd harmonic current
$I_{op.2}$ = starting current of the 2nd harmonic braking, generally taken as 40% of the calculated load current.

Criterion 2 fully uses the information from the non-fault phases and its index is the maximum value of the three phase ratios, so criterion 2 is also called maximum phase braking. The second auxiliary criterion concerning the fundamental current is used to prevent maloperation caused by a greater calculation error in smaller differential current conditions.

2.2.8.3 Criterion 3

Take the ratio of the maximum second harmonic component of the three phase currents to the maximum fundamental component of the three phase currents as an index. If the index is larger than its setting value, there is a magnetizing inrush current and the trip should be blocked.

$$I_{2.\max}/I_{1.\max} \geq K_{set} \tag{2.16}$$

Where

$I_{1.\max}$ = maximum value of the three-phase fundamental currents
$I_{2.\max}$ = maximum value of the three-phase 2nd harmonic currents.

If the inrush current of a three-phase transformer occurs, there is always a larger second harmonic current in at least one phase. Adopting the maximum second harmonic current of the three phases can achieve safe blocking and make up the shortcoming of maloperation in criterion 1 when the symmetrical inrush current arises.

2.2.8.4 Criterion 4

The second harmonic braking criteria mentioned previously only use the amplitude of the second harmonic component of the inrush current. If the phase information of the second harmonic current is added to the criterion, the brake performance on occasion of smaller second harmonic current can be improved. A new second harmonic braking criterion based on both amplitude and phase information of the inrush current has been proposed in literature (Refs [1] and [2]). First, the amplitude ratio and the phase

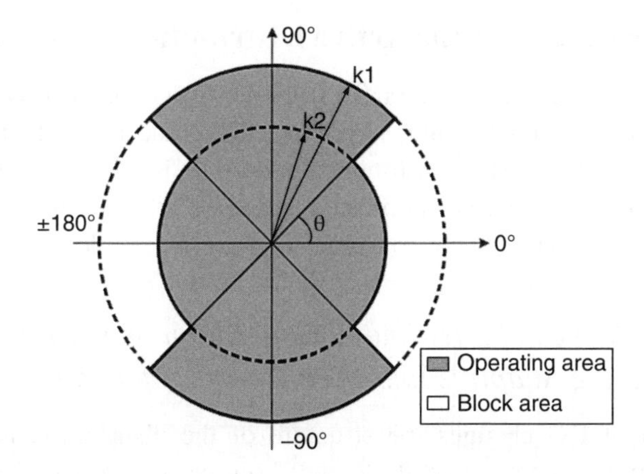

Figure 2.15 The braking area based on the amplitude ratio and the phase difference

difference of the second harmonic current to the fundamental current are defined as follows:

$$\begin{cases} k = I_2 / I_1 \\ \varphi = \arg\left(\dot{I}_2\right) - 2\arg\left(\dot{I}_1\right) \end{cases} \tag{2.17}$$

Figure 2.15 shows the scheme of second harmonic braking based on the amplitude and phase information: k1 and k2 are two threshold values of the second harmonic amplitude ratio, respectively. Usually k1 and k2 are set as 0.15 and 0.10, respectively. θ is the margin of the phase difference and usually taken as 30° [1].

The concrete criterion for second harmonic braking is as follows:

1. The current protection should be blocked if the amplitude ratio is larger than the threshold value k1.
2. The current protection should not be blocked if the amplitude ratio is less than the threshold value k2.
3. If the amplitude ratio is in the range between k2 and k1, whether to block or not is then decided by the phase difference.

Second harmonic braking is mainly used to coordinate with differential current protection of a transformer or a line powered to a transformer in order to prevent maloperation of current protection due to an inrush current from an unloaded transformer. Second harmonic braking can also be used to coordinate with three-section current protection in the medium voltage distribution network to prevent the maloperation caused by the inrush current from transformer without under-voltage protection under conditions of reclosing, especially for the instant accelerated protection trip.

2.3 Fault Protection of the Active Distribution Network

With the access of distributed generation (DG), the active distribution network has become the major trend of the grid of the future. For protection of the active distribution network, the existing protection for passive distribution network must be improved adaptively or a new wide-area protection based on artificial intelligence with the help of communication systems and smart distribution network automation devices should be established.

2.3.1 The Influence of Distributed Generation on Current Protection and the Adaptive Improvement of Protection

The integration of DG changes the structure of the distribution network from a single-ended source system to a multiple-ended source system. This further affects the value and direction of the fault current in a distribution network and causes the existing protection for passive distribution network strategy to be no longer applicable [3]. In this section, the impact of DG in different locations on the traditional current protection in distribution network is analyzed first and then some strategies to overcome these effects are introduced.

2.3.1.1 Situation 1: DG is Connected to the Distribution Bus

As shown in Fig. 2.16, DG is connected to the distribution bus. At this moment, the radial single-ended source system for each feeder has not changed. Provided that the fault occurs at the point of F1–F4, respectively, as shown in Fig. 2.16, the impact of the DG on the protection device is analyzed as follows.

1. ***Fault occurs in* F1 or F3**
   When a fault occurs in F1 or F3, both the source and the DG provide a short-circuit current to the fault point together. At this moment, comparing with the condition without

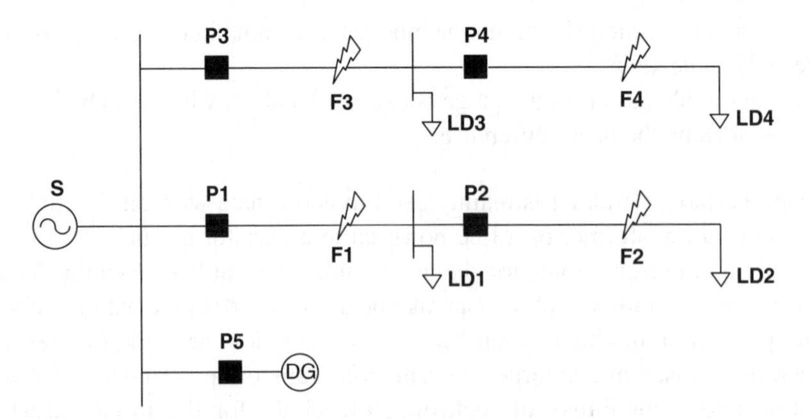

Figure 2.16 DG connected to the distribution bus

the DG, the DG increases the short circuit current flowing through P1 or P3. This increase in fault current will help improve the sensitivity of P1 or P3 and clear the fault.

2. ***Fault occurs in* F2 or F4**
When a fault occurs in F2 or F4, Both the source and the DG provide a short circuit current to the fault point together. At this moment, comparing with the condition without DG, the DG increases the short circuit current flowing through P2 or P4. This increase of the fault current will help improve the sensitivity of P2 or P4 and clear the fault.

Problem
When a fault occurs in F2 or F4, the fault current flowing through P1 or P3 is increased by the DG. If the capacity of the DG is enough to increase the fault current too much, the instantaneous over-current protection of P1 or P3 maybe be tripped incorrectly and this leads loss of protection selectivity.

Solution
The current setting of instantaneous over-current protection of P1 or P3 must be greater than the maximum short-circuit current when a fault occurs at P2 or P4. If the short-circuit current from the DG is less than 10% of the short-circuit current from the source (i.e., the stiffness of the DG connected network is no less than 10), maloperation will not occur with a reliability coefficient of 1.3. If necessary, if sensitivity is satisfied, the reliability coefficient of the instantaneous over-current protection of P1 or P3 can be increased properly.

2.3.1.2 Situation 2: DG is Connected to the Middle of the Feeder

As shown in Fig. 2.17, a DG is connected to the middle of the feeder. At this moment, the radial single-ended source system of this feeder is changed to a partly double-ended source system. Provided that a fault occurs at the point of F1–F4,

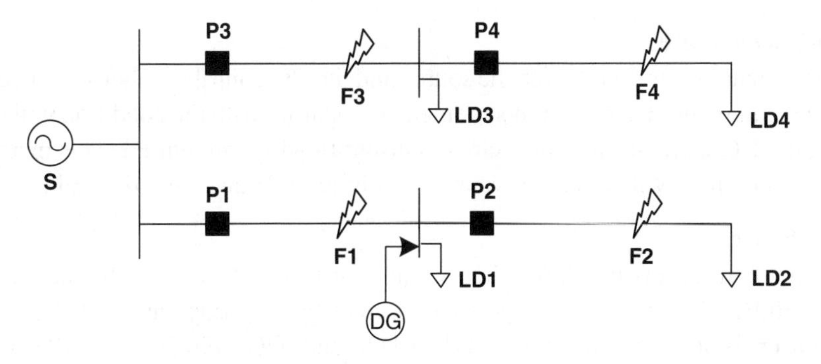

Figure 2.17 DG located in the middle of a feeder

respectively, as shown in Fig. 2.17, the impact of the DG on protection devices is analyzed as follows.

1. ***Fault occurs in* F1**

   When a fault occurs in F1, both the source and the DG provide fault current to the fault point together. But only the short circuit current from the source flows through the protection of P1 and so P1 will trip to clear the fault. It should be pointed out that the DG must be disconnected from the feeder by itself.

   Problem 1

   If a nonmetallic fault occurs in F1 and the capacity of the DG is large enough, the DG will increase the current of the fault point and also the fault voltage, this will lead to a decrease in short-circuit current provided by the source, and hence reduce the protection sensitivity of P1. Furthermore, the instantaneous over-current protection of P1 maybe refuses to trip.

   Solution 1

   A definite time over-current protection can be added as a backup protection of the instantaneous over-current protection of P1. According to the setting principle of definite time over-current protection, its sensitivity is enough to overcome the influence of the DG. Moreover, the capacity of the DG can be limited to increase the stiffness of the network.

2. ***Fault occurs in* F2**

   When a fault occurs in F2, both the source and the DG provide short-circuit current to the fault point together. At this moment, the short-circuit current from the source and the DG flows through the protection of P2. Comparing with the condition without the DG, the short-circuit current flowing through P2 is increased. Therefore, P2 is able to trip reliably and clear the fault. At the same time, although a short-circuit current also flows through P1, P2 will clear the fault prior to P1 and leaves P1 non-operational due to the coordinated time limit between P1 and P2.

3. ***Fault occurs in* F3**

   When a fault occurs in F3, both the source and the DG contribute short-circuit current to the fault point together. At this moment, comparing with the condition without the DG, the DG increases the short circuit current flowing through P3. This increase of the fault current will help improve the sensitivity of P3 and clear the fault.

   Problem 2

   When a fault occurs in F3, the short-circuit current contributed to by the DG flows through P1. This current, if large enough, may cause maloperation of P1 and form a power island composed of LD1, LD2, and DG. Finally, the DG will be disconnected.

Solution 2

An effective method is to add directional over-current protection to P1. The directional device makes P1 trip only when the short-circuit current flows from the source, and avoids the influence of inverse current from the DG. Besides, it is feasible to limit the DG capacity properly and increase the rigidity of the network.

4. ***Fault occurs in* F4**

When a fault occurs in F4, both the source and the DG contribute short-circuit current to the fault point together. At this moment, comparing with the condition without the DG, the DG increases the short circuit current flowing through P4. This increase of the fault current will help improve the sensitivity of P4 and clear the fault.

Problem 3

When a fault occurs in F4, the short-circuit current flowing through P3 is also increased compared to the condition without the DG. The increase of short-circuit current in P3 may cause a maloperation of its instantaneous over-current protection and a loss of protection selectivity.

Solution 3

It is feasible to increase the reliability coefficient of the instantaneous over-current protection of P3 properly. Besides, another effective way is to limit the DG capacity properly to improve the rigidity of the network.

Problem 4

As for Problem 2, when a fault occurs in F4, the short-circuit current contributed by the DG flows through P1. This current, if large enough, may cause maloperation of P1 and form a power island composed of LD1, LD2, and DG. Finally, the DG will be disconnected.

Solution 4

The same as Solution 2. It is feasible to add a directional over-current protection to P1. Besides, it is also effective to limit the DG capacity properly to improve the rigidity of the network.

2.3.1.3 Situation 3: DG is Connected to the End of the Feeder

As shown in Fig. 2.18, here the DG is connected to the end of the feeder. At this moment, the radial single-ended source mode of this feeder is changed to a double-ended source mode. Provided that a fault occurs at the point of F1–F4, respectively, as shown in Fig. 2.18, the impact of the DG on the protection devices is analyzed as follows.

1. ***Fault occurs in* F1**

When a fault occurs in F1, there is no fault current flowing through P3 or P4, but a large short circuit current from both the source and the DG will flow into the fault

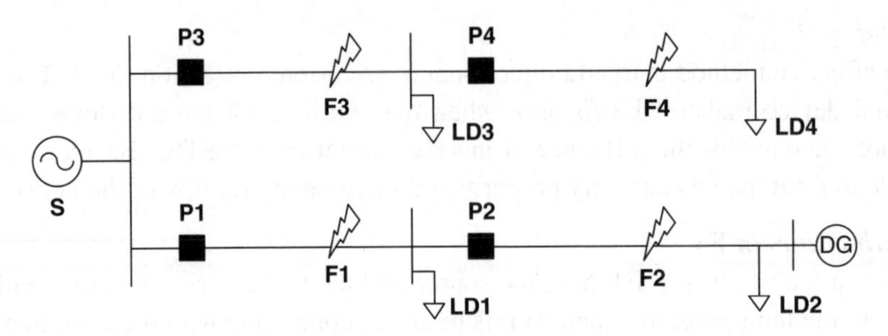

Figure 2.18 DG located in the end of feeder

point and also a large short circuit current from the source will flow through P1, a relatively small short circuit current from the DG will flow through P2.

In this condition, P1 will trip reliably and clear the fault. At the same time, if the short circuit current from DG is large enough, P2 may also trip and isolate the fault point. However, if this current is relatively small or a directional over-current protection is added to P2, the DG should disconnect by itself.

2. ***Fault occurs in* F2**
When a fault occurs in F2, there is no fault current flowing through P3 or P4, but a large short circuit current from both the source and the DG will flow into the fault point and also a large short circuit current from the source will flow through P1 and P2.

In this condition, P2 will trip reliably and clear the fault with priority, and DG should disconnect by itself to isolate the fault point.

Problem 1
If a nonmetallic fault occurs in F2 and the DG capacity is large enough to raise the voltage across the fault point, we can deduce which source the short-circuit current flowing through P1 and P2 contributed from. As a the result, protection sensitivity may decrease and the instantaneous over-current protection of P2 loses its function.

Solution 1
It is feasible to add a definite time over-current protection as the backup protection for P1. According to the settling principle of definite time over-current protection, its sensitivity should be enough to overcome this DG influence.

3. ***Fault occurs in* F3**
When a fault occurs in F3, both the source and the DG contribute short-circuit current to the fault point together. At this moment, comparing with the condition without the DG, the DG increases the short circuit current flowing through P3. This increasing of the fault current will help improve the sensitivity of P3 and clear the fault.

Problem 2
When a fault occurs in F3, the short-circuit current contributed by the DG flows through P1 and P2. This current, if large enough, may cause maloperation of P2 due

to its smaller over-current setting than P1, and thus a power island composed of LD2 and DG will be formed. Finally, the DG will disconnect by itself.

Solution 2
An effective way is to add directional over-current protection to P1 and P2. The directional device makes P1 or P2 trip only when the short-circuit current flows from the source and avoids the influence of inverse current from the DG.

4. ***Fault occurs in* F4**
When a fault occurs in F4, both the source and the DG contribute short-circuit current to the fault point together. At this moment, comparing with the condition without the DG, the DG increases the short circuit current flowing through P3 and P4. This increase of the fault current will help improve the sensitivity of P4 and P4 will clear the fault.

Problem 3
When a fault occurs in F4, the short-circuit current flowing through P3 is also increased compared to the condition without the DG. The increase of short-circuit current in P3 may cause maloperation of its instantaneous over-current protection and a loss of protection selectivity.

Solution 3
It is feasible to increase the reliability coefficient of the instantaneous over-current protection of P3 properly.

Problem 4
The same as Problem 2. When a fault occurs in F4, the short-circuit current contributed to by the DG flows through P1 and P2. This current, if large enough, may cause maloperation of P2 firstly, due to the smaller over-current setting than P1, and a power island composed of LD2 and DG will be formed. Finally, the DG will disconnect by itself.

Solution 4
The same as Solution 2. It is feasible to add directional over-current protection to P1 and P2.

This analysis shows that the influence of DG on the protection of distribution systems is closely related to the capacity of DG. If the DG capacity is very small, the influence of DG on traditional current protection can be ignored as long as the function of anti-islanding protection is adopted in the DG. But if the DG capacity is relatively large, it may lead to loss or decrease of protection sensitivity, and furthermore, may cause maloperation of the upstream protection installed in the same feeder or maloperation of protection installed in the neighbor feeders.

To sum up, decreasing DG penetration or capacity and increasing short-circuit capacity and the rigidity of the distribution system are the best strategies to solve the influence of DG on the traditional over-current protection in distribution network.

1. It is feasible to properly improve the reliability coefficient of the instantaneous over-current protection of the breaker near the source. Moreover, it is also effective to set time delay instantaneous over-current protection to make up for the decrease of the sensitivity of instantaneous over-current protection.
2. It is feasible to add a directional element to the over-current protection between the source and the DG. The protection will start only when the short-circuit current flowing from the source is detected.
3. It is also feasible to set a definite time over-current protection as a backup for all the protections between the source and the DG.
4. The grid-connected DG must have the automatic splitting functions of low-voltage, low-frequency, and anti-islanding protection.

2.3.2 Influence of Distributed Generation on Auto-Reclosure and its Adaptive Improvements

The instantaneous faults make up more than 80% of all the faults in distribution network. Due to this the use of auto-reclosure may greatly improve the reliability of the distribution system. Particularly in a radial single-ended source system, the instant accelerated protection trip with auto-reclosure has been widely used. The major influence of DG on automatic reclosing and the corresponding improvement measures are as follows.

Problem 1

DG enhances arc sustainability of the fault point. If an instantaneous fault occurs in the feeder with DG, the breakers at the source side will trip and then start the function of automatic reclosing. If DG does not disconnect from the line after the fault has occurred and remains to provide current to the fault point instead, the arc will remain in the fault point and therefore leads to the failure of automatic reclosing.

Solution 1

There are several strategies:

1. The converter-type DG must be able to detect the isolated island immediately and then be disconnected from the grid. Meanwhile, the anti-islanding protection time must be less than the reclosing time.
2. The reclosing adopts the strategies of no-voltage checking, which is to ensure reclosure starts only when the DG is disconnected and there is no voltage in the downstream network.
3. It is feasible to install a breaker with directional over-current protection and auto-reclosing at the upstream side of the DG connection point. When a fault occurs in the line, the breaker detects the short-circuit current from the DG and then trips first. After that the breaker can't be reclosed until the arc is eliminated and the

source restores its service successfully. The current setting of the directional over-current protection should be less than the rated current of the DGs in the downstream line of the protection.

Problem 2

DG may cause a non-synchronization closing of the auto-reclosing device. When an instantaneous fault occurs in the feeder with DG, the breaker at the source side will trip protectively. If the DG does not split from the grid, an isolated island will be formed. Owing to the non-synchronization between the isolated island and the grid, non-synchronized closing will have an unacceptable impact on the grid.

Solution 2

It is necessary for the ARD to have the function of synchronization checking or auto-synchronizing.

Taking the distribution network shown in Fig. 2.19 as an example, the instant accelerated protection trip of auto-reclosure is installed in P1. For the unimproved system shown in Fig. 2.19(a), if an instantaneous fault occurs in point k, P1 will trip without selectivity. If P1 recloses the breaker before the DG disconnects from the grid, reclosing will fail due to the sustainable arc. Figure 2.19(b) shows an improved system, P3 is the special added breaker with directional current protection and the function of synchronization-checking reclosing, and the current setting is less than the rated current of the DG. If an instantaneous fault occurs in point k, P1 will trip without selection, meanwhile, P3 will trip either due to the fault current from DG, then the fault will be isolated and the arc will be eliminated. P1 will automatic reclose after a time delay, thereafter, P3 will reclose synchronously when the normal source voltage is detected. Finally the power service for the whole system will be restored.

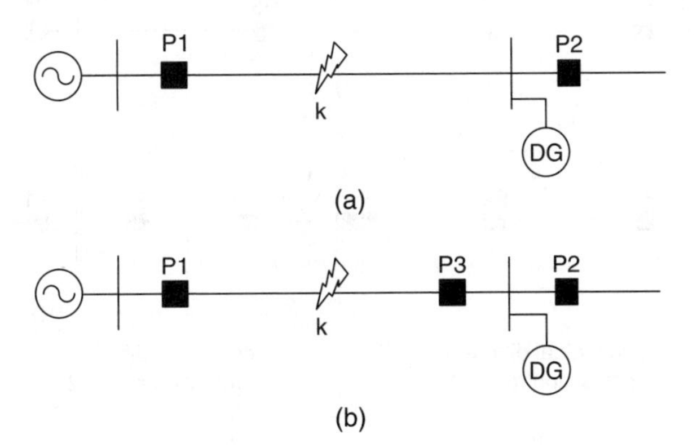

Figure 2.19 Influence of distributed generation on auto-reclosing and its improvement measure

2.3.3 Longitudinal Current Differential Protection of DG Connected Distribution Networks

This analysis shows that the adaptive improvement for traditional current protection may meet the needs of protection for the distribution network with distributed generation. However, the DG is required to have the function of anti-islanding protection, which makes it split from the grid when the system loses its source. Obviously, the DG loses its contribution to the power reliability of the distribution network and is not suitable for a micro-grid.

In a distribution network with DG or for a micro-grid, installing longitudinal current differential protection in a small area prevents the DG disconnecting from the grid. The setting principle of longitudinal current differential protection is as follows.

1. Install the protective breakers at both sides of the DG connecting point.
2. Set a longitudinal current differential protection between any two sources (including the distributed generation and the micro-grid) in the feeder.
3. The start condition of longitudinal current differential protection still adopts equation (2.12), but the minimum current that reflects the unbalanced current must consider the influence of the load in the protection range.
4. It is better to set a single shot reclosing for longitudinal current differential protection, in which the reclosing on the DG side should have the function of synchronization-checking.
5. Directional elements should be added to the definite time over-current protection in the range of the longitudinal current differential protection.

Taking the distribution network shown in Fig. 2.20 as an example, there are two DGs in the buses C and D of a feeder that can operate in island mode. According to the

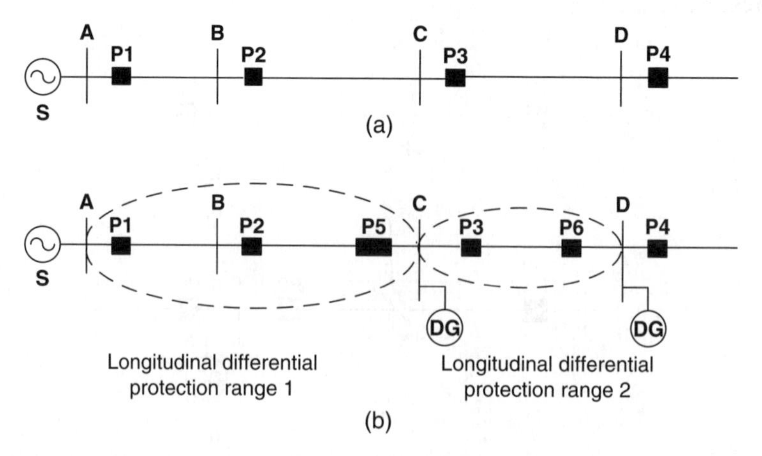

Figure 2.20 Configuration of longitudinal current differential protection in distribution grids with DG

principles mentioned here, protection at P5 and P6 is added first. Two longitudinal current differential protection configured with auto-reclosing devices are designed, respectively, in the area surrounded by P1–P5 and in the area surrounded by P3–P6.

When a fault occurs in the area 1, first the longitudinal current differential protection of P1 and P5 will clear the fault without time delay, and then the reclosure of protection P1 on the source side begins. If there is a temporary fault, the first reclosing will be successful, then protection P5 will reclose with synchronization-checking and the whole system restore power service. But if the fault is permanent, the protection will operate according to the principle of definite time over-current protection. Protection P2 will trip first if the fault occurs in the area BC, while protection P1 will trip if the fault occurs in the area AB.

When a fault occurs in area 2, first the longitudinal current differential protection of P3 and P6 will clear the fault without time delay, and then the reclosure of protection P3 on the source side begins. If there is a temporary fault, the first reclosing will be successful, then protection P6 will reclose with synchronization-checking and the whole system restores power service. But if the fault is permanent, protection P3 will trip again due to reclosing on the permanent fault.

It is worth pointing out that the current setting of the longitudinal current differential protection should be larger than the maximum magnetizing inrush current of the unloaded transformer, which is located in the differential protection area.

With the help of distributed intelligent control technology in the wide area measurement system and the distribution automation system, the longitudinal current differential protection of the distribution network with distributed generation can be promoted to a wide area current differential protection. A smart terminal unit (STU) is installed on both ends of each longitudinal differential protection; those STUs serve the functions of synchronous measurement and communication and are connected by Ethernet for data sharing. If a fault occurs somewhere in the distribution network, two adjacent STU will exchange fault current information and the protection operating signal to judge where the fault is and take appropriate action.

2.4 Coordination of Multistage Protection in the Distribution Network

The distribution automation system collects fault information from the distribution automation terminals, locates and isolates the fault according to the fault location rules, and then reconstructs the network in order to restore power service of the non-faulted area to the greatest extent. However, it will cause transient power interruption of the area in general, even if the fault only occurs in a radial branch.

Protection with coordinated relaying can clear a fault rapidly and restore the power supply to the non-faulted area without power interruption. However, the coordinated protection relaying in the distribution network is often difficult to achieve. On a practical level, the override and multiple stage trips usually happen due to the unreasonable configuration of multiple stage protection relaying and unreasonable parameter settings.

For the open-loop operating urban distribution network with short power supply radius but more segments, the short-circuit current in each segment switch makes little difference. Generally, only the cooperative scheme of protection relaying with pure delay time difference can be used to realize that the branch fault cleared will not influence trunk lines and the sub-branch fault cleared will not influence the branch. However, by the delay time differential coordination, coordinating multistage protection relaying in the distribution trunk line cannot be achieved.

For the suburban or rural distribution network with a long power supply radius, there are obvious differences in the fault current in each segment switch when a fault occurs in the trunk line. At this time, it is feasible to adopt multistage coordination of three-section over-current protection.

This section discusses the possible coordination patterns, characteristics, and setting methods of multistage protection relaying in the distribution network.

2.4.1 Time Difference Based Coordination of Multistage Protection in the Distribution Network

For the open-loop operating mode of an urban distribution network with a short power supply radius but more segments, the short-circuit current only makes a little difference when the fault occurs in the different place. Hence, it is impossible to set different current setting values for different switches. At this time, the only way to clear the fault selectively is by the coordination of the delay time differential in multistage protections.

This section mainly discusses the multistage protection coordination based on the time differential, that is, the protection coordination is implemented by setting different delay times for the 10 kV outlet switch and 10 kV feeder switch in the transformer substation.

2.4.1.1 Feasibility Analysis

2.4.1.1.1 Case without Instantaneous Over-Current Protection

Instantaneous over-current protection can be uninstalled, but instead of time delay instantaneous over-current protection or definite-time over-current protection, in such cases where a line fault cannot cause bus voltage of power station or an important user getting lower than the 60% of rated voltage, the conductor section of the line is large enough to allow the fault to clear with a time delay, and the time limit of the over-current protection is less than 0.5~0.7 s. The time delay instantaneous over-current protection or definite-time over-current protection makes the coordination of multistage differential protection possible.

2.4.1.1.2 *Case with Instantaneous Over-Current Protection*

A 10 kV feeder operating in the open-loop mode can be regarded as a single-ended source network, the current setting of instantaneous over-current protection (section I) $I_{\mathrm{I.op}}$ should be larger than the maximum three-phase short-circuit current at the end of the line,

$$I_{\mathrm{I.op}} \geq K_{\mathrm{rel}}^{\mathrm{I}} I_{\mathrm{k.max}}^{(3)} \tag{2.18}$$

Where, $K_{\mathrm{rel}}^{\mathrm{I}}$ is the reliability coefficient and $K_{\mathrm{rel}}^{\mathrm{I}} \geq 1.3$, $I_{\mathrm{k.max}}^{(3)}$ is the maximum three-phase short-circuit current at the end of the line, which can be expressed as

$$I_{\mathrm{k.max}}^{(3)} = \frac{\dot{E}}{\mathrm{j}X_{\mathrm{s.min}} + \sum_{i=1}^{n} \dot{Z}_i} \tag{2.19}$$

Where

$\dot{E}$ = equivalent potential of system source
$X_{S.\min}$ = equivalent reactance of the system source in the maximum operation mode
$Z_1, Z_2, \ldots, Z_n$ = feeder impedance of each line segment from the head to the end of the feeder.

Assuming that feeders have the same conductor section, then

$$I_{\mathrm{k.max}}^{(3)} = \frac{\dot{E}}{\mathrm{j}X_{\mathrm{s.min}} + l_{\mathrm{total}} \cdot \dot{z}} \tag{2.20}$$

Where, l_{total} is the total length of the feeder and z is the feeder impedance per unit length.

The main phase to phase fault that occurs in the overhead line distribution network is the two-phase short-circuit. Although most phase to phase faults in the cable can become three-phase faults gradually, the possibility of cable faults is very low. Faults, mostly two-phase faults, usually occur in the branch or user side of overhead line and the two-phase short-circuit current $I_{\mathrm{k}}^{(2)}$ is 0.866 times the three-phase fault current. The critical length l_{C} on the feeder in the upstream of the fault point, at which the two-phase short-circuit in the maximum operation mode doesn't cause the operation of the instantaneous over-current protection, is

$$\left|\frac{0.866\dot{E}}{\mathrm{j}X_{S.\min} + l_C \cdot \dot{z}}\right| = 1.3\left|\frac{\dot{E}}{\mathrm{j}X_{S.\min} + l_{\mathrm{total}} \cdot \dot{z}}\right| \tag{2.21}$$

And this formula can be rewritten as

$$\frac{\left(X_{S.\min} + l_{\mathrm{total}} x_0\right)^2 + l_{\mathrm{total}}{}^2 r_0^2}{\left(X_{S.\min} + l_C x_0\right)^2 + l_C^2 r_0^2} = 2.26 \tag{2.22}$$

Table 2.1 The ratio of the critical length to the whole length of the feeder (l_{total} = 10 km)

Type	Wire model	$l_C/l_{total}(\%)$		
		$X_{S.min}=1\Omega$	$X_{S.min}=0.5\Omega$	$X_{S.min}=0.2\Omega$
Overhead line	LGJ-150	59.52	63.12	65.19
	LGJ-185	58.95	62.82	65.06
	LGJ-240	58.29	62.46	64.91
Cable	YJV-150	50.36	59.90	64.28
	YJV-185	46.47	58.09	63.61
	YJV-240	41.93	55.78	62.71
	YJV-300	38.53	53.93	61.95

Table 2.2 The ratio of the critical length to the whole length of the feeder (l_{total} = 5 km)

Type	Wire model	$l_C/l_{total}(\%)$		
		$X_{S.min}=1\Omega$	$X_{S.min}=0.5\Omega$	$X_{S.min}=0.2\Omega$
Overhead line	LGJ-150	51.88	59.52	63.82
	LGJ-185	50.86	58.95	63.57
	LGJ-240	49.70	58.29	63.29
Cable	YJV-150	24.30	50.36	61.48
	YJV-185	16.40	46.47	60.07
	YJV-240	8.67	41.93	58.22
	YJV-300	3.29	38.53	56.73

Where, r_0 is the resistance per unit length and x_0 is the reactance per unit length of the feeder.

According to Equation (2.22), under different system reactance $X_{s.min}$ and wires or cable types, the ratios of critical feeder length to whole feeder length, l_C/l_{total}, are shown in Tables 2.1–2.3.

The feeder equipped with instantaneous over-current protection is divided into two parts by the critical length l_C. When a two-phase short circuit occurs in the upstream part, instantaneous over-current protection of the substation outlet switch will trip and this case does not have the coordination conditions of multistage differential protection. But if a two-phase short circuit occurs in the downstream part, the instantaneous over-current protection of the substation outlet switch will not trip and then over-current protection with time delay will start. In other words, the downstream part of the critical point l_C on the feeder possesses the coordination conditions of multistage differential protection, as shown in Fig. 2.21.

It should be noted that the critical length l_C is calculated in the maximum operation mode of the system. In the mode of non-maximum operation, the critical length point

Table 2.3 The ratio of the critical length to the whole length of the feeder (l_{total} = 2 km)

Type	Wire model	$l_C/l_{total}(\%)$		
		$X_{S.\min}=1\Omega$	$X_{S.\min}=0.5\Omega$	$X_{S.\min}=0.2\Omega.$
Overhead line	LGJ-150	26.88	47.89	59.52
	LGJ-185	25.03	46.68	58.95
	LGJ-240	22.88	45.31	58.29
Cable	YJV-150	0	8.03	50.36
	YJV-185	0	0	46.47
	YJV-240	0	0	41.93
	YJV-300	0	0	38.53

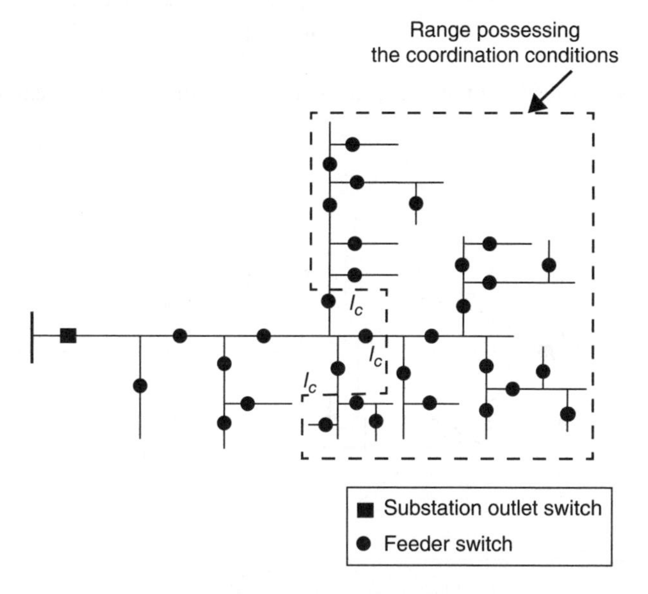

Figure 2.21 The range possessing the multistage differential protection coordination conditions

on the feeder will shift toward upstream, such as point l_C' shown in Fig. 2.22. This shift is beneficial for the coordination of multistage differential protection in the downstream part of the feeder.

The 10 kV feeders are often emitted from the main substation and the power range shows a fan-like distribution under normal circumstances. The closer to downstream, the more branches there are, as shown in Fig. 2.23. Therefore, for the feeder installed with instantaneous over-current protection, the area processing the coordination conditions of multistage differential protection just lies in the downstream area with more branches. This condition is beneficial for the coordination of multistage differential protection of substation outlet switch, branch switch, and/or sub-branch switch (or user switch).

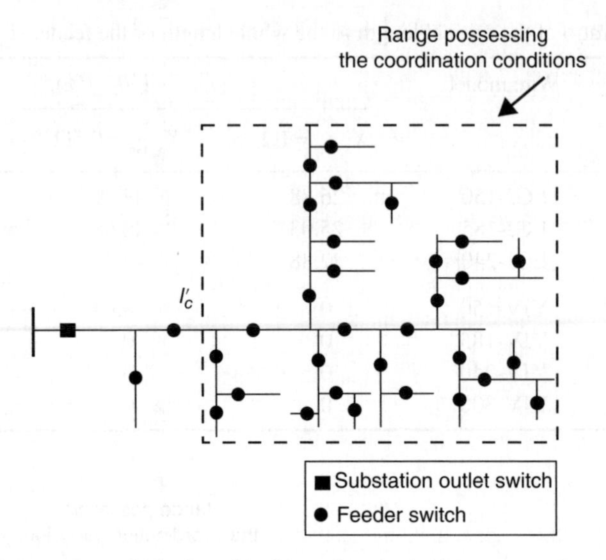

Figure 2.22 The upstream shift of a critical length point in the non-maximum mode operation of the system

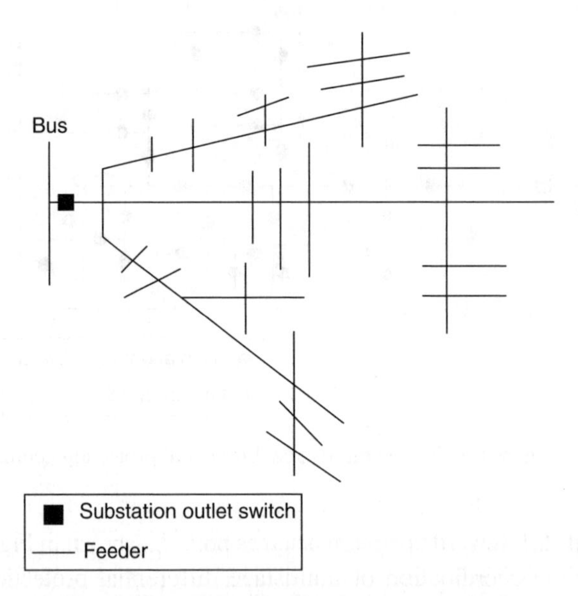

Figure 2.23 Fan-like power range of the feeder

2.4.1.1.3 The Setting of the Time Differential

The analysis here shows that the coordination of multistage differential protection can be implemented along the whole length of the feeder without instantaneous over-current protection, and the coordination of multistage differential protection can be implemented in the downstream part of the feeder with instantaneous over-current protection.

The operating time of the over-current protection of the transformer secondary switch in the substation (i.e., 10 kV bus feeding switch) is generally set to 0.5~0.7 s. Considering the worst case, in order not to influence the time setting of upper stage protection, it is necessary to arrange the multistage differential protection time delay coordination at intervals of 0.5 s.

For the feeder's circuit breaker with a spring-energy-storage operation mechanism, mechanical operation time is generally 60~80 ms and the inherent response time of protection is about 30 ms. Considering a certain time margin, the time differential ΔT can be set to 250~300 ms to achieve coordination of two-stage differential protection.

For the feeder's circuit breaker with a permanent magnet operation mechanism, its opening time is about 20 ms. A fast protection algorithm can complete fault diagnosis in about 10 ms. Considering a certain time margin, the time differential ΔT can be set to 150~200 ms to achieve the coordination of three-stage differential protection.

If the system has the ability to withstand a greater short-circuit current, the time delay of the over-current protection in the transformer secondary switch can be extended properly in order to improve reliability of multistage differential protection coordination. For example, for the switch with a permanent magnetic operation mechanism, the time differential can be set to 200 ms. For the switch with a spring-energy-storage operation mechanism, the time differential can be set to 300 ms.

Generally, the thermal stability check-time of transformer, breaker, load switch, disconnector, line, and the current transformer is 2s. Therefore, the coordination project of multistage differential protection proposed here does not influence the thermal stability of these devices.

2.4.1.2 Configuration Rules for Multistage Protection based on Time Differential

2.4.1.2.1 Configuration Rules for Two-Stage Differential Protection

For the coordination of two-stage differential protection, the principles of protection configuration and switch choice are as follows.

1. All trunk line switches adopt the load switch.
2. User switch, branch switch, and sub-branch switch adopt the circuit breaker.
3. The instantaneous over-current protection of the substation outlet switch can be installed as needed, and time delay of the over-current protection is set to one time differential, ΔT.
4. For the area possessing multistage differential protection coordination conditions, time delay of over-current protection in the user breaker or branch breaker is set to 0 s. The current setting should be greater than the maximum load current and magnetizing inrush current from unloaded transformer.

The advantages of the two-stage differential protection configuration mentioned here can be seen as follows.

1. When a fault occurs in the branch or user area (or sub-branch), the corresponding branch breaker or user breaker will trip first while the operation of substation outlet switch is not needed. Therefore, this configuration avoids power failure of the whole line and reduces the number of the users who suffer a power failure after the fault.
2. Prevents the multistage switches from tripping at the same time or overrides tripping. Therefore, this configuration is a simple fault processing procedure, there is a shorter switch operation time, and short restoration time for a temporary fault.
3. The use of load switch reduces the cost compared to the breaker.

2.4.1.2.2 Configuration Principle of Three-Stage Differential Protection

A typical three-stage differential protection is formed by the 10 kV substation outlet switch, the branch switch, and user switch in the area possessing coordination multistage differential protection conditions. Time delay of the over-current protection in the user switch is set to 0 s and the current setting should be larger than the maximum load current and magnetizing inrush current of the unloaded transformer. The time delay of over-current protection in the branch switch is set to one time differential, ΔT, and the current setting should be larger than the maximum load current and magnetizing inrush current of unloaded transformer. The time delay of the over-current protection in the substation outlet switch is set to dual time differential, $2\Delta T$.

2.4.1.3 The Influence of Grid-Connected Distributed Generation

The integration of the distributed generation (DG) may have some impact on the scheme of the multistage differential protection, as mentioned previously.

2.4.1.3.1 The Influence on the Feeder without Instantaneous Over-Current Protection

For a feeder without instantaneous over-current protection, its multistage differential protection coordination is relatively simple, which can be realized by the coordination of the time delay of over-current protection in the substation outlet switch and branch switch or user switch.

After the access of DG, in the case of a phase to phase fault, the short-circuit current flowing through the substation outlet breaker and each segment switch along the feeder is changed compared with that before the access of DG. On the one hand, when a fault occurs downstream of the DG, it will decrease the short-circuit current flowing through the switch in the upstream of the same feeder, and increase the short-circuit current flowing through the downstream switch. When a fault occurs upstream of the DG, a short-circuit current from only DG will flow through the switches located between the fault point and the DG in the same feeder. On the other hand, the DG in the other feeders can also inject a short-circuit current to this feeder through the same distribution bus, which may lead to a rise of short-circuit current flowing through the substation outlet breaker in the fault feeder and the segment switches at the upstream of the fault point. A short-circuit current contributed to by the DG in the fault-free feeder will flow through its corresponding feeder's outlet breaker. As for the actual increase or decrease of the short-circuit current somewhere in the feeder, it is necessary to comprehensively analyze the impact of these two aspects.

However, because sensitivity of over-current protection is generally high, the DG capacity is generally small, and in particular, the short-circuit current from the inverter-based DG is very small, so the DG will not cause maloperation of over-current protection in the upstream switch.

So, for the feeder without instantaneous over-current protection, the access of the DG generally has no influence on its multistage differential protection coordination.

2.4.1.3.2 The Influence on the Feeder with Instantaneous Over-Current Protection

Provided that a feeder is installed with instantaneous over-current protection, and a DG is connected to an adjacent feeder powered by the same distribution bus. When a fault occurs on this feeder, the current from the DG in the adjacent feeder will expand the protection range of instantaneous over-current protection in the substation outlet switch and its operating zone will extend to downstream of the critical length l_C, which is calculated according to the case with no DG, resulting in the increased possibility of an override trip. Therefore, in this case, the area with multistage differential protection coordination needs to be properly adjusted to the downstream.

The DG does not affect instantaneous over-current protection in the same feeder. If a fault occurs downstream of the DG, the short-circuit current flowing through the substation outlet switch will be decreased, and the protection range of instantaneous over-current protection will be reduced, which is beneficial for the multistage differential protection coordination between substation outlet switch and downstream switch of critical length point l_C.

2.4.1.3.3 Impact of the Reverse Short-Circuit Current from DG on the Multistage Differential Protection

In addition, no matter whether instantaneous over-current protection is installed or not, if a DG provides a larger short-circuit current, it is likely that the over-current protection without directional element of the branch or user switch, and even the substation outlet switch, would not operate properly due to the reverse fault current, and result in a failure of multistage differential protection coordination. For this, it is necessary to selectively configure a directional element in the corresponding switch. However, when the short-circuit current from DG is lower (such as the inverter-based DG, photovoltaic generation, etc.), it has less of an effect in this regard and if necessary, it is feasible to appropriately adjust the current setting of over-current protection rather than installing a directional element.

2.4.1.3.4 The Influence on Auto-Reclosing

According to the "technical rule for the DG connected to power grid" formulated by the State Grid Corporation of China, it is proposed that the non-islanding DG must be removed from the grid 2s after faulting in the feeder. Therefore, for the switch with a downstream DG, the auto-reclosing delay time should be more than 2s.

2.4.1.4 Handling Measurement of the Area without Multistage Differential Protection Coordination

For the feeder with instantaneous over-current protection, the upstream part of the critical length point l_c does not meet the coordination condition for multistage differential protection. For this part of the feeder, it is feasible to adopt a similar configuration to that described in Section 2.4.1.2. That is to say, the substation outlet switch, branch switch, and user switch (or sub-branch switch) are replaced by circuit breakers with instantaneous over-current protection. Although an override trip may occur when a fault occurs, it is still beneficial for fault processing with the help of coordination of auto-reclosing, or centralized intelligent fault processing from the distribution automation system's master station. For specific fault processing, refer to Chapter 3.

2.4.2 The Coordination of Multistage Protection Based on Three-Section Over-Current Protection in the Distribution Network

Protection relaying with a differential coordination for distribution network cannot achieve the multistage protection coordination for the trunk line. For suburban or rural distribution lines with a longer power radius, when a fault occurs in the trunk line, there is an obvious difference of short-circuit current in each segment switch located upstream of the fault, and it is feasible to coordinate the multistage three-section over-current protection.

This section mainly discusses the key problems and improved methods of the coordination of multistage three-section over-current protection for a distribution network.

2.4.2.1 Analysis of the Coordination of Multistage Three-Section Over-Current Protection under the Traditional Setting Method

According to the "Setting guide for 3~110 kV power system protection equipment (DL/ T584–2007)" formulated by the State Grid Corporation of China, the current setting of instantaneous over-current protection (section I) of the traditional three-section

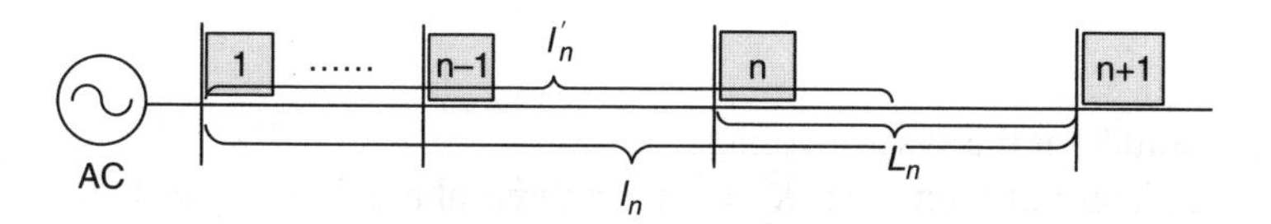

Figure 2.24 Schematic of configurable *n*-stage protection

over-current protection should be set according to the principle that this protection should not operate when a maximum three-phase short-circuit occurs at the end of the protection zone. The current setting of time delay instantaneous over-current protection (section II) should be set according to the principle of protecting the whole zone at this stage in any fault, and it also should coordinate with section I of the next stage, while sensitivity must be checked according to the minimum two-phase short-circuit current at the end of this stage. The current setting of definite time over-current protection (section III) should be set to avoid protection operation at the maximum load current, so the setting value is relatively small. Section III protection should not only protect the whole length at this stage, but also play the role of remote backup protection in order to protect the whole length of the next stage.

As for the feeder shown in Fig. 2.24, it is assumed that in-stage protection coordination can be achieved. That is to say, the feeder is divided into *n* segments, and there are *n* sets of three-section over-current protection device along the line. The path consists of three-section over-current protection devices at each stage and is called the protection path of the feeder.

Where l_n represents the length from a 10 kV substation bus to the end of the nth segment in the protection path, and its equivalent impedance is denoted by Z_n, l_n' represents the length from 10 kV substation bus to the protection end of section I on the *n*th segment in the protection path, and its equivalent impedance is denoted by Z_n'. Let L_n denote the length of the *n*th segment, it can be expressed as

$$L_n = l_n - l_{n-1} \quad \left(\mathrm{n} > 0, 1_0 = 0\right) \tag{2.23}$$

Provided that system impedance in the maximum and minimum mode is, respectively, denoted by $X_{S.min}$ and $X_{S.max}$, for the *n*th-stage instantaneous over-current protection (section I), in any operation mode in which the system impedance is X_s ($X_{s.min} < X_s < X_{s.max}$), when a fault occurs at the end of the zone of the instantaneous over-current protection, the short-circuit current $I_{k.n}$ shall be equal to the setting of the nth stage instantaneous over-current protection (section I) $I_{I.op}$,

$$I_{k.n} = I_{I.op} \tag{2.24}$$

That is

$$\frac{K_\varphi U}{\sqrt{3}\left|X_s + Z'_n\right|} = \frac{K^{I}_{rel} U}{\sqrt{3}\left|X_{s.min} + Z_n\right|} \tag{2.25}$$

Where

U = the amplitude of the system voltage.
K_φ = the coefficient of fault type, $K_\varphi = 1$ for a three-phase fault, $K_\varphi = 0.866$ for a two-phase fault.
$K^{\mathrm{I}}_{\mathrm{rel}}$ = reliability coefficient of the protection, usually $K^{\mathrm{I}}_{\mathrm{rel}} = 1.3$.

Let β denote the length ratio of a stage that the instantaneous over-current protection of this stage protection device must protect, That is, when a two-phase short circuit occurs in the minimum operation mode of the system, the length ratio that this segment of the line needs to have to be protected is at least β, there are

$$\begin{cases} \dfrac{l'_n - l_{n-1}}{l_n - l_{n-1}} \geq \beta \\ \dfrac{0.866.U}{\sqrt{3}\left|X_{s.\max} + Z'_n\right|} = \dfrac{K^{\mathrm{I}}_{\mathrm{rel}}U}{\sqrt{3}\left|X_{s.\min} + Z_n\right|} \end{cases} \tag{2.26}$$

And it can be rewritten as

$$\begin{cases} \dfrac{l'_n - l_{n-1}}{l_n - l_{n-1}} \geq \beta \\ \left|X_{s.\min} + Z_n\right| = \dfrac{K^{\mathrm{I}}_{\mathrm{rel}}}{0.866}\left|X_{s.\max} + Z'_n\right| \end{cases} \tag{2.27}$$

In order to meet the requirements of reliability coordination, the range of time delay instantaneous over-current protection (section II) of the $(n-1)$th stage should be less than the instantaneous over-current protection (*Section* I) of the nth stage, and its sensitivity K_{sen} must meet the requirements, namely

$$I^{(2)}_{\mathrm{k}.n-1.\min} \geq K_{\mathrm{sen}} I_{\mathrm{II.op}.n-1} \tag{2.28}$$

And it can be rewritten as

$$\left|X_{\mathrm{s.min}} + Z_n\right| \geq \frac{K_\Sigma}{0.866}\left|X_{\mathrm{s.max}} + Z_{n-1}\right| n = 2,3\ldots\ldots \tag{2.29}$$

Where

K_Σ = equivalent coefficient by comprehensively converting, $K_\Sigma = K_{\mathrm{sen}} K^{\mathrm{II}}_{\mathrm{rel}} K^{\mathrm{I}}_{\mathrm{rel}}$
K_{sen} = sensitivity coefficient, generally it is no less than 1.3
$I^{(2)}_{\mathrm{k}.n-1.\min}$ = the minimum two-phase short circuit current at the end of the $(n-1)$th stage of the feeder
$I_{\mathrm{II.op}.n-1}$ = the protection setting value of section II of the $(n-1)$th stage.

Both the formulas (2.27) and (2.29) are the general conditions to achieve the coordination of three-section over-current protection at the nth stage.

Provided that the wire type and its section of the trunk line is the same, and the impedance per unit length is represented by z, then

$$Z_i = l_i z \tag{2.30}$$

Put Equation (2.30) into (2.27) and take a critical value, then the following formula can be obtained,

$$\frac{l_n^2 r_0^2 + \left(X_{\text{s.min}} + l_n x_0\right)^2}{\left[\beta l_n + (1+\beta) l_{n-1}\right]^2 r_0^2 + \left[X_{\text{s.max}} + \left(\beta l_n + (1-\beta) l_{n-1}\right) x_0\right]^2} = \left(\frac{K_{\text{rel}}^{\text{I}}}{0.866}\right)^2 \tag{2.31}$$

And put Equation (2.30) into (2.29) and take a critical value, then the following formula can be derived,

$$\frac{l_n^2 r_0^2 + \left(X_{\text{s.min}} + l_n x_0\right)^2}{l_{n-1}^2 r_0^2 + \left(X_{\text{s.max}} + l_{n-1} x_0\right)^2} = \left(\frac{K_\Sigma}{0.866}\right)^2, n \geq 2 \tag{2.32}$$

Where, r_0 and x_0 are the resistance and the reactance per unit length of the feeder, respectively.

By Equations (2.31) and (2.32), the protection range of l_n can be calculated separately, and the larger of these two formulas is the minimum length needed to achieve three-section over-current protection coordination of the nth stage.

2.4.2.2 Analysis of Improved Multistage Protection Coordination based on the Differentiated Settings

The short-circuit type is not considered in the setting of the instantaneous over-current protection (section I) of the traditional three-section over-current protection. The current value is set according to the principle that it should be larger than the maximum short-circuit current of the three-phase short-circuit at the end of the line and sensitivity is checked according to the minimum two-phase short-circuit current. Thus, when a two-phase short circuit occurs, the protection range is relatively small. If the line is very short or the system operation mode varies widely, the protection range may be zero, this will lead to loss of protection function. In particular, in the distribution network with an overhead line, the two-phase fault occurs in most cases.

In fact, protection relaying devices can easily distinguish the three-phase short-circuit from the two-phase short-circuit. If the three-phase short circuit and two-phase short circuit are treated separately, the setting value of the instantaneous over-current protection is set with the maximum fault current at the end of the line, and the sensitivity is checked with the minimum fault current, respectively, two different settings can be calculated for the two different faults. Then, in the case of two-phase short circuit, the protection range will be much larger than that obtained from conventional methods. Therefore, the line length or the system operation mode will have little effect on the protection range and the protection reliability can be greatly improved.

Using the approach similar to Section 2.4.2.1, the general conditions that are needed for configuring the *n*th-stage improved three-section over-current protection based on the current setting difference between three-phase short-circuit and two-phase short-circuit, can be deduced as follows.

$$\begin{cases} \dfrac{l'_n - l_{n-1}}{l_n - l_{n-1}} \geq \beta \\ \left|X_{s.\min} + Z_n\right| = K^{\mathrm{I}}_{\mathrm{rel}} \left|X_{s.\max} + Z'_n\right| \\ \left|\left|X_{s.\min} + Z_n\right|\right| \geq K_{\Sigma} \left|\left|X_{s.\max} + Z_{n-1}\right|\right| \end{cases} \tag{2.33}$$

Under the conditions that the wire type and section of the trunk feeder are all the same, the minimum length l_n needed for this improved n-stage protection coordination based on the current setting difference as before can be obtained by the following Equation (2.34).

$$\begin{cases} \dfrac{l_n^2 r_0^2 + \left(X_{\mathrm{s.min}} + l_n x_0\right)^2}{\left(\beta l_n\right)^2 + r_0^2 + \left(X_{\mathrm{s.max}} + \beta l_n x_0\right)^2} = \left(K^{\mathrm{I}}_{\mathrm{rel}}\right), n = 1 \\ \dfrac{l_n^{\,2} r_0^{\,2} + \left(X_{\mathrm{s.min}} + l_n x_0\right)^2}{l_{n-1}^2 r_0^{\,2} + \left(X_{\mathrm{s.max}} + l_{n-1} x_0\right)^2} = \left(K_{\Sigma}\right)^2, n \geq 2 \end{cases} \tag{2.34}$$

According to this analysis, we can draw out the critical surface of each stage of protection at different system capacity. In the following figures, the system capacity in the minimum operation mode is $S_{\min}$ = 200~500 MVA, the system capacity in the maximum operation mode is $S_{\max}$ = 500~1000 MVA, the overhead line is LGJ-240, the geometric distance is 2.5 m, the reliability coefficients are $K^{\mathrm{II}}_{\mathrm{rel}}$ = 1.2 and $K^{\mathrm{I}}_{\mathrm{rel}}$ = 1.3, respectively, the sensitivity coefficient is K_{sen} = 1.5, and β is 20%. Then in the area above the critical surface, configuration of multistage protection is possible, while in the area below the surface, configuration is impossible. Figure 2.25 is the critical surface of the four-stage protection under the traditional setting method, while Fig. 2.26 is the critical surface of four-stage protection under the improved coordination method.

The comparison of Figs 2.25 and 2.26 can help us draw the conclusion that when the system capacity and the power radius are certain, more stage can be configured by the improved setting method based on the fault type than the traditional method, and the instantaneous over-current protection range is greatly increased in the case of a two-phase short circuit.

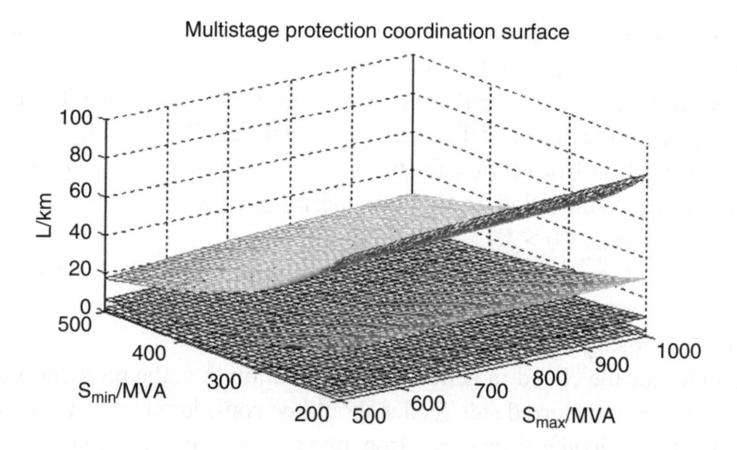

Figure 2.25 Critical surface of four-stage protection coordination under the traditional setting method

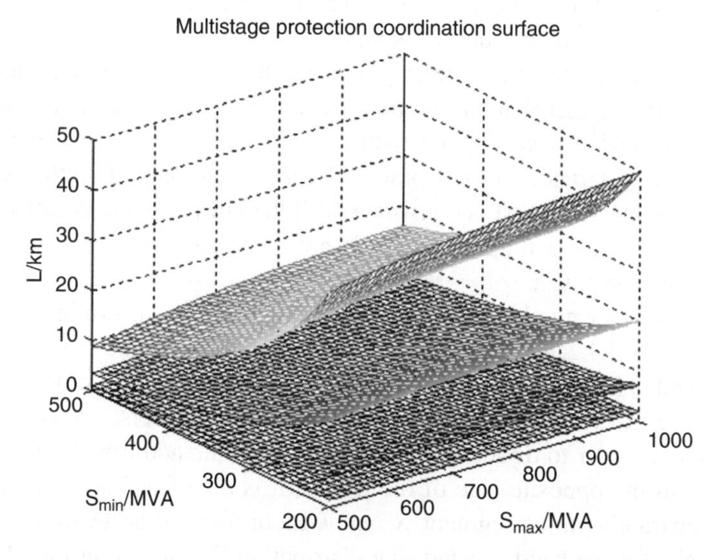

Figure 2.26 Critical surface of four-stage protection coordination under the improved coordination method

2.4.2.3 Configuration of Multistage Three-Section Over-Current Protection

It is feasible to calculate the desired minimum length l_n (i.e., the lower limit of the location of $(n + 1)$th protection device) of the feeder to configure the n-stage three-section over-current protections according to Sections 2.4.2.1–2.4.2.2. Then, when the power radius L is certain, the maximum stage number of the three-section over-current protection, which can be configured in this feeder, can be determined.

If $l_n = L$, only nth-stage protection can be installed. In fact, considering an appropriate margin, this condition can be rewritten as $1.05 > L/l_n > 1.0$.

If $l_n < L < l_{n+1}$ (considering an appropriate margin, such as: $1.0 > L / l_{n+1}$ and $L / l_n > 1.05$), the $(n + 1)$ stage protections can be installed. Because the protection range of this stage can't meet the conventional length requirement for a $(n + 1)$ stage, so the $(n + 1)$th stage of protection is called an "additional stage". Generally the additional stage only sets section I protection for the whole length of just this stage, section II protection is usually uninstalled. In fact, considering an appropriate margin, this condition can be rewritten as $1.0 > L/l_{n+1}$ and $L/l_n > 1.05$.

For the multi-source distribution network with a tie line to the opposite side, the stage number of protection should be determined by considering the situation when the tie switch is closing.

For example, for the "hand in hand" ring distribution line, the protection coordination problem of the transferred side feeder should be considered when the tie switch is closed and the feeder load is transferred from one feeder to another feeder to realize the multistage protection. After the tie switch is closed, the power flow direction of the feeder that is transferred has changed, and the relationship between upstream and downstream flow from the switch along the line has changed, so the original properly set current settings are not appropriate anymore.

To solve these problems, it is necessary to install the directional power elements for the hand in hand ring distribution line to achieve multistage protection coordination. Each protection device is configured with the fault power directional element and, according to the difference in fault power direction, two sets of setting values are adopted. The setting value that is consistent with the normal mode is called the "forward setting", while the setting value that is not consistent with the normal mode is called the "reverse setting".

The protection stage number of the two directions should be determined by the principles in Sections 2.4.2.1–2.4.2.2 under the consideration of the two hand-in-hand feeders comprehensively and the condition that the tie switch is in closed mode. If a protection device doesn't need reverse protection, block the reverse protection directly.

Sometimes, in order to divide the feeder evenly by the additional reverse protection relaying in the opposite side of the feeder, it is also feasible to set up reverse protection separately on the segment switch without forward protection.

For example, for the hand in hand ring distribution line shown in Fig. 2.27(a), S1 and S2 are outlet breakers of the 10 kV substation, A, B, C, E, and F are segment breakers, and D is the tie switch. In this figure, the solid body represents a switch in a closing state, while the hollow body represents a switch in an opening state; a square represents a breaker, while a circle represents a load switch. This assumes that the feeder of S1 configures four-stage forward three-section over-current protection and the feeder of S2 configures three-stage forward three-section over-current protection. According to the analysis, in the condition that the load on feeder S2 is transferred to S1, the reverse protection is set at point F if five-stage protection coordination can be achieved, as shown in Fig. 2.27(b). Also according to the analysis, in the condition that the load on feeder S1 is transferred to S2, the reverse protection is set at point B if four-stage

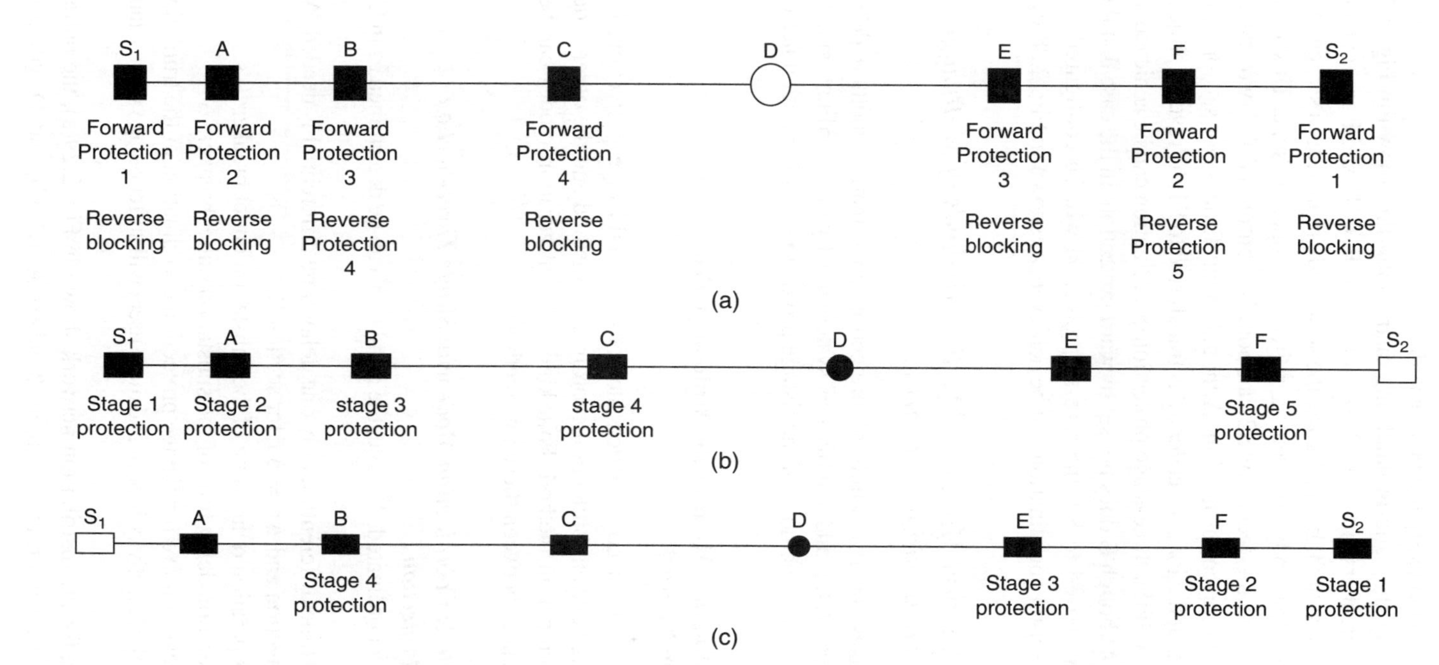

Figure 2.27 First example of multistage protection configuration in a "hand-in-hand" ring network

protection coordination can be achieved, as shown in Fig. 2.27(c). The reverse protection in the remaining switches should be blocked.

What's more, for the hand-in-hand ring distribution line shown in Fig. 2.28(a), S1 and S2 are outlet breakers of the 10 kV substation, A, B, C, E, and F are segment breakers, and D is the tie switch. In this figure, the solid body represents a switch in a closing state, while the hollow body represents a switch in an opening state. This assumes that the feeder of S1 and the feeder of S2 can both be configured for two-stage forward three-section over-current protection. Although point E does not set forward protection, in the case that the load of feeder S2 is transferred to S1 the reverse protection will be configured at E if the three-stage protection coordination can be achieved, as shown in Fig. 2.28(b). Although B does not set forward protection, in the case that the load of feeder S1 is transferred to S2, the reverse protection will be configured at B if the three-stage protection coordination can be achieved, as shown in Fig. 2.28(c).

2.4.3 Coordination Modes and Setting Methods of Multistage Protection of Distribution Networks

This section takes the two kinds of protection coordination method described in Sections 2.4.1 and 2.4.2, and summarizes systematically the coordination mode, characteristics, and setting methods of multistage protection for distribution networks.

2.4.3.1 Coordination Mode of the Multistage Protection for Distribution Network

There are two kinds of coordination methods for multistage protection, one is the three-section over-current protection coordination method and the other is delay time differential coordination method. Four kinds of configuration modes can be derived from the differences between these two methods.

2.4.3.1.1 Mode 1: Coordination Mode with Single Three-Section Over-Current Protection

Stages possibly coordinated: N stage, the number N can be determined as in Chapter 1.

Required delay time differential: only one delay time differential is needed, $\Delta t_I = 0$ for section I protection and $\Delta t_{II} = \Delta T$ for section II.

The minimum operation delay time of the substation outlet breaker: 0 s.

The longest operation delay time of the substation outlet breaker: ΔT.

Advantages: it can achieve multistage protection coordination of the trunk line.

Disadvantages: the selectivity is poor and more users will suffer outage when a fault occurs.

For example, for the distribution network shown in Fig. 2.29(a), the outlet breaker S1 of a 10 kV substation and segment breakers A, (B, C), and D will constitute a

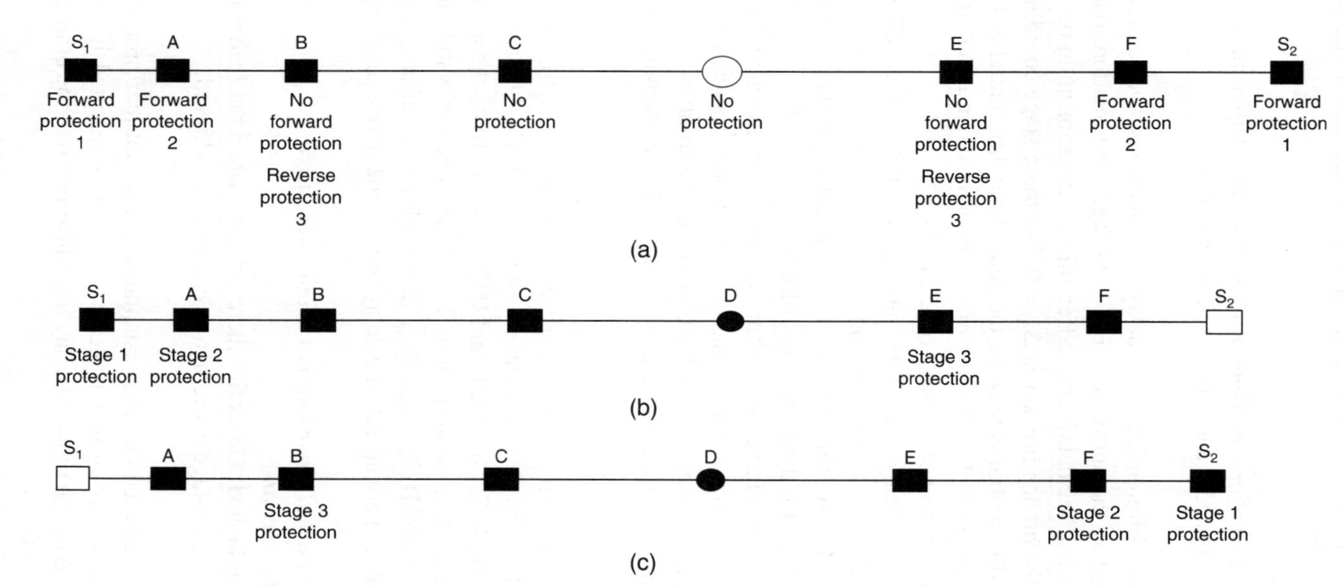

Figure 2.28 Second example of multistage protection configuration in a "hand-in-hand" ring network

four-stage three-section over-current protection, and B and C are both in the third stage. The time delays of Section I protection on the substation outlet breaker S1 and segment breakers of A, B, C, and D are all 0 s, while the time delay of section II protection is ΔT.

2.4.3.1.2 Mode 2: Coordination Mode with Single Time Differential Protection

Stages possibly coordinated: generally, not more than three.

Required delay time differential: time delay instantaneous over-current protection should be adopted in the substation. For two-stage coordination, one time delay differential is needed, the delay time differential of the substation outlet breaker is $\Delta t_I = \Delta T$, while the branch breaker is $\Delta t_0 = 0$. For three-stage coordination, a two-stage time delay differential is needed, the delay time differential of the substation outlet breaker is $\Delta t_2 = 2\Delta T$, while the branch breaker is $\Delta t_I = \Delta T$, the delay time differential of the sub-branch or user breaker is $\Delta t_0 = 0$.

The minimum operation delay time of the substation outlet breaker: ΔT for two-stage coordination and $2\Delta T$ for three-stage coordination.

The longest operation delay time of the substation outlet breaker: ΔT for two-stage coordination and $2\Delta T$ for three-stage coordination.

Advantages: two-phase and three-phase short-circuit can coordinate fully, in which the branch fault does not influence the trunk line, and the sub-branch/user fault does not influence the branch. Fewer users will suffer power interruption.

Disadvantages: The substation must adopt time delay instantaneous over-current protection.

For example, for the distribution network shown in Fig. 2.29(b), the outlet breaker S1 of the 10 kV substation, branch breakers (A1, B1, B2, C1, D1, and D2), and sub-branch breakers (A11, A12, C12, D21, and D22) will constitute the three-stage delay time differential protection coordination. The delay times of sub-branch breakers (A11, A12, C12, D21, and D22) are all 0 s, and the delay times of branch breakers (A1, B1, B2, C1, D1, and D2) are all ΔT: the delay time of the outlet breaker S1 is $2\Delta T$.

2.4.3.1.3 Mode 3: Partial Coordination Mode with Single Time Differential Protection

Stages possibly coordinated: Generally there are two, when the feeder is long or the wire section is thin, three-stage coordination can also be achieved.

Required delay time differential: The instantaneous over-current protection and time delay instantaneous over-current protection should be adopted in the substation, $\Delta t_I = 0$ and $\Delta t_{II} = \Delta T$. The required delay time differential of the branch/sub-branch/user switch is $\Delta t_0 = 0$.

The minimum operation delay time of the substation outlet breaker: 0 s.

The longest operation delay time of the substation outlet breaker: ΔT.

Advantages: Instantaneous over-current protection can be used in the substation, the branch fault does not influence the trunk line, and fewer users will suffer power interruption.

Disadvantages: When the feeder is short or the wire section is thick, the coordination can only be achieved in part of two-phase short circuit.

For example, for the distribution network shown in Fig. 2.29(c), when the two-phase short circuit fault occurs in downstream of the branch breaker (B2, C1, C2, D1, and D2), these branch breakers and the outlet breaker S1 of a 10 kV substation will constitute a two-stage delay time differential protection coordination, of which the delay time of the branch breaker (B2, C1, C2, D1, and D2) and the outlet breaker S1 of a 10 kV substation is 0 s.

2.4.3.1.4 Mode 4: The Mixed Coordination Mode of Three-Section Over-Current Protection and the Delay Time Differential Protection

The trunk line adopts the three-section over-current protection, but all the branch and sub-branch lines adopt delay time differential protection.

Stages possibly coordinated: $n + 2$ stages for a fully coordinated mode, and $n + 1$ stages for a partial coordination mode.

Required delay time differential:

1. When it is mixed with the two-stage fully coordinated delay time differential protection, the required delay time differential of section I of the substation outlet breaker and the trunk line breaker is $\Delta t_{I} = \Delta T$, the required delay time differential of *Section* II of the substation outlet breaker and the trunk line breaker is $\Delta t_{II} = 2\Delta T$. The required delay time differential of the branch/sub-branch/users breaker is $\Delta t_{0} = 0$.
2. When it is mixed with the three-stage fully coordinated delay tine differential protection, the required delay time differential of section I of the substation outlet breaker and the trunk line switch is $\Delta t_{I} = 2\Delta T$, the required delay time differential of section II of the substation outlet breaker and the trunk line switch is $\Delta t_{II} = 3\Delta T$, $\Delta t_{I} = \Delta T$ for the branch breaker, and $\Delta t_{0} = 0$ for the sub-branch/user breaker.
3. When it is mixed with partial coordination two-stage delay time differential, the required delay time differential of section I of the substation outlet breaker and the trunk line breaker $\Delta t_{I} = 0$, and the required delay time differential of section II of the substation outlet breaker and the trunk line breaker is $\Delta t_{II} = \Delta T$, the required delay time differential of branch/sub-branch/user breaker is $\Delta t_{0} = 0$.

The minimum operation delay time of the substation outlet breaker:

- The case mixed with the two-stage full coordination: ΔT
- The case mixed with the three-stage full coordination: $2\Delta T$
- The case mixed with the two-stage partial coordination: 0.

The maximum operation delay time of the substation outlet breaker:

- The case mixed with the two-stage full coordination: $2\Delta T$
- The case mixed with the three-stage full coordination: $3\Delta T$
- The case mixed with the two-stage partial coordination: ΔT.

Advantages: The selectivity is enhanced and fewer users will suffer a power interruption.

Disadvantages: When it is mixed with the full coordination, the protection rapidity of the substation outlet switch will decrease. When it is mixed with the partial coordination, the selectivity can be improved for a part of two-phase fault.

For example, for the distribution network shown in Fig. 2.29(d), the outlet breaker S1 of a 10 kV substation and segment switches A, B, and C will constitute four-stage three-section over-current protection coordination. The delay time of section I protection of the substation outlet breaker S1 and the segment breakers of A, B, and C is $2\Delta T$, and the delay time of section II protection of the substation outlet breaker S1 and the segment breakers of A, B, and C is $3\Delta T$. S1, A, B, C, branch breakers (A1, B1, C1, D1, and D2), and sub-branch breakers (A11, B21, D11, and D12) will constitute a three-stage fully coordinated delay time differential protection. The delay time of sub-branch breakers (A11, B21, D11, and D12) is 0 s, and the delay time of branch breakers (A1, B1, C1, D1, and D2) is ΔT. This example describes the mixed mode of the three-section over-current protection and the three-stage fully coordinated delay time differential protection.

As another example, for the distribution network shown in Fig. 2.29(e), the outlet breaker S1 of a 10 kV substation and segment switches A, B, and C will constitute four-stage three-section over-current protection coordination. The delay time of Section I protection of the substation outlet breaker S1 and segment breakers of A, B, and C is 0 s, and the delay time of section II protection of the substation outlet breaker S1 and segment breakers of A, B, and C is ΔT. The delay time of branch breakers (B2, C1, D1, and D2) is 0 s. When a two-phase short circuit occurs downstream of branch breakers (B2, C1, D1, and D2), S1, A, B, C, and branch breakers (B2, C1, D1, and D2) will constitute a two-stage partial coordinated delay time differential protection. This example describes the mixed mode of the three-section over-current protection and two-stage partial coordinated delay time differential protection.

2.4.3.2 Setting Principle of the Multistage Over-Current Protection in a Distribution Network

The setting principle of delay time of the coordinated multistage protection in a distribution network has been discussed in the previous sections; this section only discusses the setting principle of current of coordinated multistage protection in distribution network. According to the "Setting guide for 3~110k V power system protection equipment" (DL/T584–2007), the current setting can be determined by the following principles.

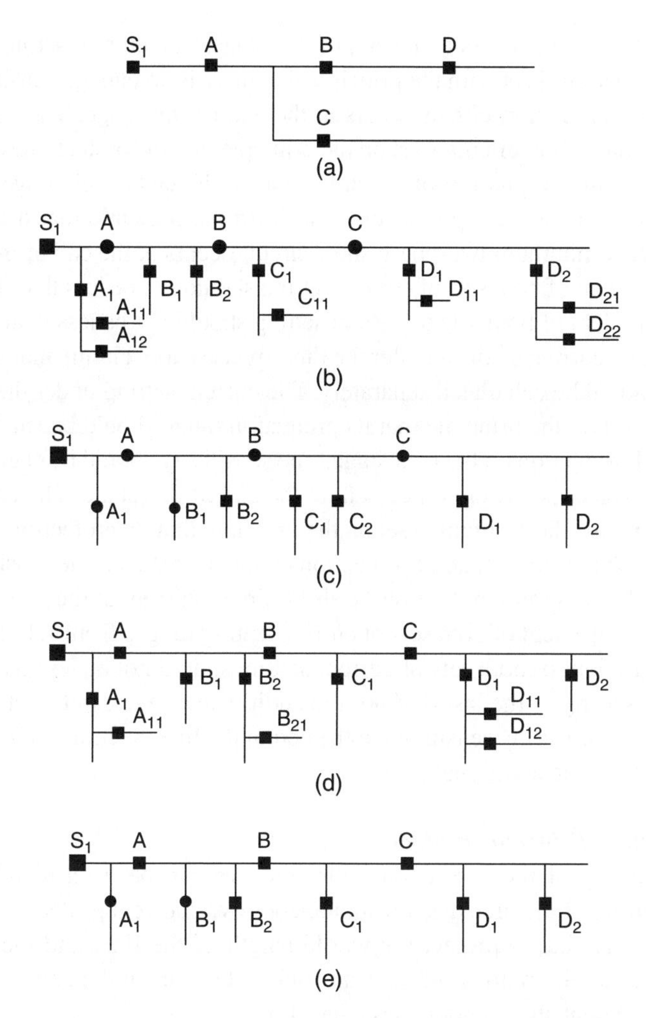

Figure 2.29 Typical coordination modes of multistage protection in distribution network

It should be noted that the current setting only concerns with the line itself for a radial line. For a distribution network with multiple sources, the current setting needs comprehensive consideration in the case that the tie switch is closing and on one side the feeder powers additional loads transferred from the other source side. For example, for the hand-in-hand ring network, the determination of current setting should comprehensively consider both feeders and is at the condition of the switch closing.

2.4.3.2.1 Setting Principle of Mode 1

Section I: The current settings under the three-phase short-circuit and the two-phase short-circuit should be calculated separately. The current setting under the three-phase short-circuit is set on the principle that the protection should not trip when the maximum three-phase short-circuit occurs at the end of this stage. The current setting of the additional stage protection is set with the principle that there is an enough sensitivity when a minimum three-phase short-circuit occurs at the end of this stage. The current setting under the two-phase short circuit is set on the principle that the protection should not trip when the maximum two-phase short-circuit occurs at the end of this stage. The current setting of the additional stage protection is set with the principle that there is enough sensitivity when a minimum two-phase short-circuit occurs at the end of this stage. The reliability coefficient of both sets of current setting should not be less than 1.3, while the sensitivity coefficient of both sets of current setting should not be less than 1.5.

Section II: The current settings under the three-phase short-circuit and the two-phase short-circuit should be calculated separately. The current setting under the three-phase short-circuit is set on the principle that its protection range should be smaller than that of the Section I protection on the next stage. The sensitivity should be checked with the minimum three-phase short-circuit current at the end of this stage. The current setting under the two-phase short-circuit is set on the principle that its protection range should be smaller than that of the Section I protection on the next stage. The sensitivity should be checked with the minimum two-phase short-circuit current at the end of this stage. The reliability coefficient of both sets of current setting ranges from 1.1~1.2, while the sensitivity coefficient of both sets of current setting should not be less than 1.5.

The current setting of the last (but not the additional) stage can be set according to the requirement of ensuring sensitivity at the end of the line. Section II protection is not usually installed in the additional stage.

2.4.3.2.2 Setting Principle of Mode 2

The current setting of the substation outlet breaker can be determined as follows. The current setting of the time delay instantaneous over-current protection is set on the principle that it is able to protect the whole length of the line, and there is enough sensitivity when a minimum two-phase short-circuit occurs at the end of the line. The sensitivity coefficient should not be less than 1.5.

The current setting of the branch/sub-branch/user breaker can be determined as follows. The current setting is set on the principle that the protection should not trip when the maximum load current and magnetizing inrush current from downstream unloaded transformer flows through the protection device. The reliability coefficient should not be less than 1.5.

2.4.3.2.3 Setting Principle of Mode 3

The current setting of the substation outlet breaker can be determined as follows.

Section I: The current setting is set on the principle that the protection should not trip when the maximum three-phase short-circuit occurs at the end of this stage. The reliability coefficient should not be less than 1.3.

Section II: The current setting is set on the principle that it is able to protect the whole length of the line and there is enough sensitivity when a two-phase fault occurs at the end of the line in the minimum operation mode of the system. The sensitivity coefficient should not be less than 1.5. For a radial line, the current setting only concerts with the line itself. For a hand in hand ring network, the current setting concerts with the both feeders and should be at the condition of the tie switch closing.

The current setting of the branch/sub-branch/user breaker can be determined as follows. The current setting is set on the principle that the protection should not trip when the maximum load current and magnetizing inrush current from the downstream unloaded transformer flows through the protection device. The reliability coefficient should not be less than 1.5.

2.4.3.2.4 Setting Principle of Mode 4

1. *N + 2 for full coordination*

   The current setting of the trunk line breaker should be set on the principle as same as section I and section II in mode 1.

   The current setting of the branch/sub-branch/user breaker should be set on the principle that the protection should not trip when the maximum load current and magnetizing inrush current from downstream unloaded transformer flows through the protection device. The reliability coefficient should not be less than 1.5.
2. *N + 1 for partial coordination*

   The current setting of the trunk line breaker can be determined as follows.

   Section I: The current setting is set in the traditional way. That is to say, the current setting is set on the principle that the protection should not trip when the maximum three-phase short-circuit occurs at the end of this stage. The current setting of the additional stage protection is set on the principle that there is an enough sensitivity when a minimum two-phase short-circuit occurs at the end of this stage. The reliability coefficient should not be less than 1.3, while the sensitivity coefficient should not be less than 1.5.

   Section II: The current setting is set in the traditional way. That is to say, the current setting is set on the principle that it is able to protect the whole length of this stage in any case but its protection range should be smaller than that of Section I protection on the next stage. The sensitivity should be checked with the minimum two-phase short-circuit current at the end of this stage. The reliability coefficient of both sets of current setting ranges from 1.1 to 1.2, while the sensitivity coefficient of both sets of current setting should not be less than 1.5.

   The current setting of the last but not the additional stage is set to ensure enough sensitivity. Section II protection is not usually installed in the additional stage.

   The current setting of the branch/sub-branch/user breaker should be set on the principle that the protection should not trip when the maximum load current and magnetizing inrush current from a downstream unloaded transformer flows through the protection device. The reliability coefficient should not be less than 1.5.

There are some other issues should be noted in the following.

1. For a radial branch, if an additional stage is added, the branch lines on the additional stage can't achieve partial coordination, due to the fact that Section I can protect the whole length of this stage and hence Section II protection is not installed. Therefore, the additional stage would be uninstalled if the trunk line of the additional stage is short and there are more branch lines.
2. For an interconnected distribution network, the current setting is concerned with the operation mode of the tie feeders.

2.4.3.2.5 Inrush Current Problem

The influence of magnetizing inrush current during the energizing of a distribution transformer on protection relay in a distribution network is enormous. Without considering the remanence, the closing inrush current of a distribution transformer can reach 5–8 times its rated current in the worst case. The inrush current can be even larger with the consideration of remanence. Generally, in the three-phase transformer with a phase difference of 120° in three-phase voltage, there is always one phase in the worst condition and the maximum magnetizing inrush current exists in this phase.

The closing inrush current that flows through the feeder switches is quite different from the magnetizing inrush current of a single transformer and it is a superimposition of the magnetizing inrush current of all distribution transformers downstream of this switch. When a fault occurs, a breaker with a protection device will trip to clear the fault and this action will lead to the power failure of each distribution transformer downstream of the breaker at the same time. Therefore, the remanence will be roughly of the same polarity. The closing afterwards (e.g., from a reclosure) will make each transformer energize again at the same time. Therefore, the amplitude of the closing inrush current that flows through this switch is a superimposition of the magnetizing inrush current of each distribution transformer.

For the distribution transformer, the inrush current is enormous at the closing moment. However, the inrush current will decrease rapidly to a negligible range by 7–10 cycles (i.e., 0.14~0.2 s). Therefore, the closing inrush current has little effect on the section II protection (time delay instantaneous over-current protection) and section III protection (definite time over-current protection), but it has an enormous effect on section I protection (instantaneous over-current protection). If the current setting of section I protection is smaller than the inrush current, the closure to restore power service will not succeed due to the protection malfunction. For the case where the breaker configures a single-shot reclosure, the inrush current is the main cause of the reclosing failure for an instantaneous fault. In this case, great attention will often be paid to seek the "permanent fault" that did not originally exist, and hence leads to a long-term power failure.

The main ways to solve the inrush current problem include:

1. Adopt a second harmonic restraint or detect the intermittent angle to distinguish the inrush current and the fault current, avoiding maloperation of section I protection.

2. Sacrifice the sensitivity of section I protection and increase the current setting to a larger value than the magnetizing inrush current. For the breaker with non-delay protection (such as section I) on the branch, due to a relatively small transformer capacity of downstream of the branch, the sympathetic inrush current that flows through the branch breaker is generally small, so it is possible to avoid the inrush current influence by sacrificing protection sensitivity.
3. Sacrifice the rapidity of section I protection by a time delay of 0.1~0.15 s. By this time, most of the inrush current has already decreased.
4. Restore the power service stage by stage: This measure can decrease the number of the distribution transformers that produces the sympathetic inrush current during every closing, thus decreasing the influence of the inrush current.
5. Cancel the delayed accelerated protection and add an appropriate time delay in reclosing control to decrease the influence of the inrush current.

2.4.3.2.6 Influence of DG

The influence of DG on the multistage coordination of time differential protection has been discussed in Section 2.4.1.3. This section mainly discussed the influence of DG on the multistage coordination of three-section over-current protection.

As mentioned, for the distribution network with DG, if a phase-to-phase short-circuit occurs, the short-circuit current that flows through the substation outlet breaker and each segment switch along the feeder has changed compared to the case without a DG. Those changes need to be analyzed comprehensively when considering DG location on this feeder or the adjacent feeder from the same bus.

When the DG capacity is small and the DG is a converter type, the influence of it on the short-circuit current is generally small or negligible. The multistage configuration and the current setting principle of three-section over-current protection discussed in Section 2.4.3.3 can also be used.

When the DG capacity is enormous or the DG is a motor type, generally the influence on the short-circuit current cannot be ignored. The influence on the multistage configuration and the setting principle of three-section over-current protection discussed in Section 2.4.3.3 needs to be analyzed. The reverse protection can be canceled, but for forward protection, one of the following two measures should be adopted.

1. Adopt a more strict setting principle. It is necessary to appropriately increase the reliability coefficient and the sensitivity coefficient. First, determine the current setting again without consideration of the DG, then check its reliability and sensitivity with the DG. The setting can be used if the check is passed.
2. If the reliability and sensitivity check is not passed, the coordination method of the auto-reclosing and grid-disconnected characteristics of photovoltaic generation should be adopted as follows:
   a. The configuration of three-section over-current protection on the substation outlet breaker is the same as described in Section 2.4.2, but a delay time of the single-shot auto-reclosure should be extended to 2~3 s.

b. The over current pulse counting function is set in the three-section over-current protection along the feeder and the starting threshold is set to one appearance of over-current or loss of voltage. That is to say, when the protection device experiences over-current for the first time, the three-section over-current protection is blocked. Only after the reclosing of the substation outlet breaker, will the second experience of over-current start up the three-section over-current protection.

With this configuration, when a fault occurs, the protection of the substation outlet breaker will trip (at this moment, the other switches along the feeder are maintained at the closing state), and all DGs will split from the grid within 2 s due to loss of voltage. After a time delay more than 2~3 s, the substation outlet circuit breaker will be reclosed. If the fault is temporary, the closing will succeed and the DG will return to the grid and restore to the normal state gradually. If the fault is permanent, the three-section over-current protection of the corresponding breaker along the feeder will trip under protection coordination.

2.4.4 *Example Analysis*

Figure 2.30 shows the 10 kV distribution network of a village. In this figure, the outlet line of the substation named X1 appears as a complex branch and is divided into two trunk lines. One is the No. 1 trunk line from the substation X1 to the tie switch N, its length is 11 km and uses the overhead line LGJ-240. Another is the No. 2 trunk line from X1 to the greenhouse Y for vegetable, its length is 27 km and uses overhead line LGJ-150. The No. 3 trunk line comes from the substation named X2 to the tie switch N, its length is 8 km and it also uses the overhead line of LGJ-240. The other lines are branch lines. Switches A–H are the segment breakers on the trunk lines. Switches A_i–H_i are branch or user breakers. It is assumed that the short-circuit capacity of the system in the maximum operation mode is 400 MVA and the short-circuit capacity of the system in the minimum operation mode is 260 MVA. The reliability coefficient $K^{II}_{rel} = 1.2$ and $K^{I}_{rel} = 1.3$, the sensitivity coefficient $K_{sen} = 1.5$, and $\beta = 20\%$.

To fully illustrate the method discussed here, the protection relaying is configured in the following two situations.

2.4.4.1 Situation 1: Provided that the Current Protection on the Substation Outlet Breaker can be Delayed by a Time of ΔT

According to the description in Section 2.4.3, it is appropriate to adopt the $N + 2$ full coordination mode in mode 4. The trunk line adopts three-section over-current protection, but the branch line adopts protection coordinated by the delay time differential.

For the trunk line 1 and the trunk line 3, because the two trunk lines contact each other via the tie switch, the configuration should be considered comprehensively at the

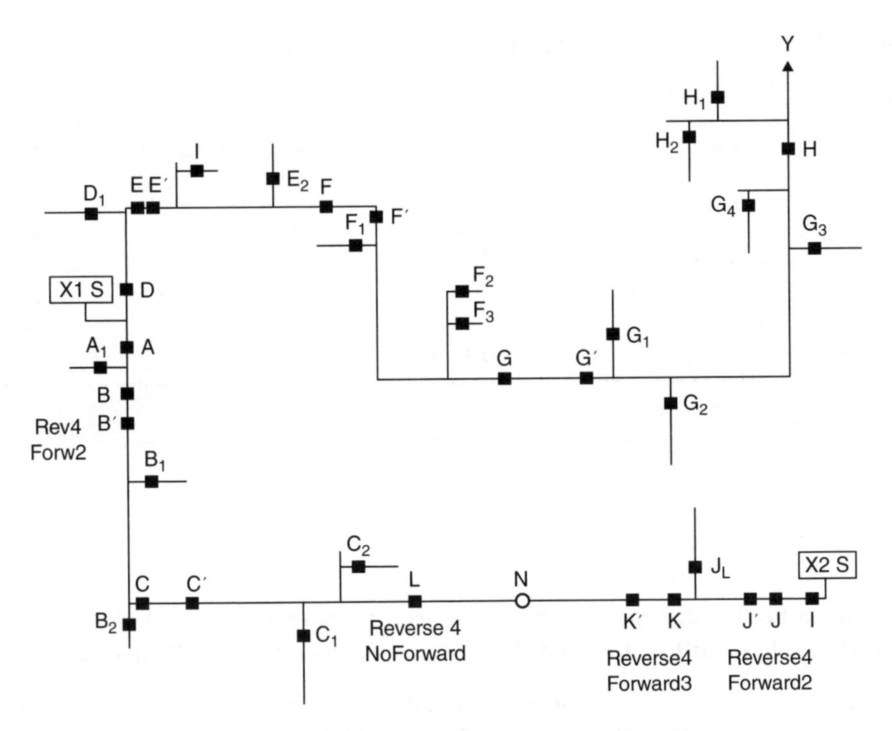

Figure 2.30 A 10 kV distribution network of a village

condition of the tie switch closing (i.e., the load of one trunk line can be transferred to the other). The trunk line 1 can be configured with three-stage forward protection located in A, B, and C, respectively. The third protection can be extended to the trunk line 3 and an additional stage with reverse protection can be located at K. Trunk line 3 can be configured with three-stage forward protection located in I, J, and K, respectively. The third protection can be extended to the trunk line 1, and an additional stage with reverse protection can be located at L.

The trunk line 2 is a radial feeder and can be configured with five-stage three-section over-current protection located at D, E, F, G, and H; of these H is the additional stage of the protection.

It is known from the short-circuit calculation that the minimum three-phase short-circuit current of each branch is 430 A and the minimum two-phase short-circuit current is 372 A. Therefore, the current setting of all the branch breakers can be set to 280 A for a three-phase fault and 240 A for a two-phase fault. The time delay is set to 0 s. Adopting the improved multistage protection coordination based on the differences of settings, the current setting of each stage protection on the three trunk lines is shown in the Table 2.4. In this table, the first set of setting values is used for the three-phase short-circuit and the second set of setting values is used for the two-phase short-circuit.

Table 2.4 Current settings of each protection on the trunk lines in situation 1

Protection stage	Trunk Line 1		Trunk Line 2		Trunk Line 3	
	Section I setting/kA	Section II setting/kA	Section I setting/kA	Section II setting/kA	Section I setting/kA	Section II setting/kA
Stage 1	12.75/11.04	5.33/4.6	12.78/11.07	5.34/4.62	12.75/11.04	5.33/4.6
Stage 2	4.44/3.84	2.1/1.82	4.45/3.85	2.11/1.82	4.44/3.84	2.1/1.82
Stage 3	1.24/1.08	0.62/0.54	1.76/1.52	0.87/0.75	1.75/1.52	0.87/0.75
Stage 4	[0.43]	—	0.73/0.63	0.36/0.32	[0.43]	—
Stage 5	—	—	0.32/0.28	—	—	—

Notes:

1. Values in front of the slash are the first set of setting values, while values after the slash are the second set of setting values.
2. Values in the [ ] are reverse setting values, others are forward setting values.

2.4.4.2 Situation 2: Provided that the Current Protection on the Substation Outlet Breaker can't be Delayed (Instantaneous Over-Current Protection)

According to the description in Section 2.4.3, it is appropriate to adopt the $N + 1$ partial coordination mode of mode 4. The trunk line adopts the three-section over-current protection, but the partial branch line adopts the protection coordinated by the delay time differential.

For trunk line 1 and trunk line 3, because the two trunk lines contact each other via the tie switch, the configuration should be considered comprehensively at the condition of the tie switch closing (i.e., the load of one trunk line can be transferred to the other). The trunk line 1 can be configured with three-stage forward protection located in A, B′, and C′, respectively. The third protection can be extended to the trunk line 3 and an additional stage with reverse protection can be located at J′. Trunk line 3 can be configured with three-stage forward protection located in I, J′, and K′, respectively. The third protection can be extended to the trunk line 1 and an additional stage with reverse protection can be located at B′.

Trunk line 2 is a radial feeder and can be configured with four-stage forward three-section over-current protection located at D, E′, F′, and G′; of these G′ is the additional stage of the protection. The current setting of each stage protection on the three trunk lines can be set by the traditional methods of three-section over-current protection and the setting values are shown in Table 2.5.

It is known from the short-circuit calculation that the minimum two-phase short-circuit current is 372 A. Therefore, the current setting of part of the branch breakers (A_1, B_2, J_1, D_1, E_2, F_2, F_3, G_3, and G_4) can be set to 240 A and the time delay is set to 0 s. Certainly, the current setting of each branch breaker can also be set with a different short-circuit current.

Table 2.5 Current settings of each protection on the trunk lines in situation 2

Protection stage	Trunk Line 1		Trunk Line 2		Trunk Line 3	
	Section I setting/kA	Section II setting /kA	Section I setting/kA	Section II setting/kA	Section I setting/kA	Section II setting/kA
Stage 1	10.45	3.91	10.43	3.9	10.45	3.91
Stage 2	3.26	1.36	3.25	1.37	3.26	1.36
Stage 3	1.07	0.46	1.14	0.49	1.07	0.46
Stage 4	[0.43]	—	0.28	—	[0.43]	—

Note: values in the [ ] are reverse setting values, others are forward setting values.

2.5 Summary

1. The distribution network with a grid-connected DG becomes an active network and the traditional protection method that was originally used for the single-ended source system is no longer suitable. The improved scheme and new protection method is proposed.
2. According to the situations where the DG is located in different places in the distribution network and with different configurations of reclosing, this chapter proposes the corresponding adaptive improved strategies in the field of current protection configuration, ARDs, and DG. It is worth pointing out that wide-area differential protection technology based on the distribution intelligence control technology of a distribution automation system is the future developmental direction for active distribution networks.
3. For the substation outlet breaker that has not got instantaneous over-current protection installed, the three-stage differential protection coordination on the user (sub-branch), branch, and substation outlet breakers can be realized. The fault in the branch cannot influence trunk lines and faults from users (sub-branch) cannot influence the branch either. The coordination of multistage protection relaying can improve the ability to clear faults in distribution automation systems.
4. Because most of phase-to-phase faults are two-phase faults, for the substation outlet breaker with instantaneous over-current protection, partial feeders still can coordinate by stage differential. So, coordination of protection relaying can improve the ability to clear a fault in the distribution automation system.
5. For the feeder uninstalling instantaneous over-current protection, the connection of DG will not affect the coordination of multistage differential protection in general. For the feeder installing instantaneous over-current protection, if the DG is accessed to the adjacent feeder by the same bus, instantaneous over-current protection range of the substation outlet breaker will extend, but the coordination range of the multistage differential protection will decrease. In order to avoid maloperation of the

over-current protection due to a larger reverse fault current provided by the DG in this feeder, it is necessary to selectively configure the directional element.

6. For a distribution network with a long powering radius, the multistage three-section over-current protection coordination can be realized if there is a obvious difference in the short-circuit current along the line. The improved setting method based on the difference between the setting values of three- and two-phase faults can improve the sensitivity and protection range in a two-phase fault under the traditional setting method.
7. There are four-coordination modes, the coordination mode with single three-section over-current protection, the mode with single time differential protection, the partial coordination mode with simple delay time differential protection, and the mixed coordination mode with three-section over-current protection and the delay time differential protection, these modes are all derived from synthesizing the delay time differential and three-section over-current protection coordination methods. The characteristics of the four modes are analyzed, the configuration method and the setting principle of protection parameters are proposed.

3

Fault Processing Based on Distributed Intelligence

Liu Jian, Xu Shiming and Chen Xingying

Abstract
Fault processing approaches based on devices with distributed intelligence are described. Two types of automatic switch coordination methodologies without communication are described, such as, FA based on recloser and voltage-delay type sectionalizers and FA based on reclosing with a fast over-current protection mode. The fast healing approach based on neighbor communication is also dealt with.

Keywords
feeder automation (FA), fault processing approaches, distributed intelligence, temporary fault, permanent fault, fault isolation, service restoration, reclosers, voltage-delay type sectionalizers, reclosing with fast over-current protection, fast healing approach based on neighbor communication

3.1 Introduction

This chapter discusses fault processing based on devices with distributed intelligence, including two types of automatic switch coordination methodologies: one without communication and one requiring a high speed, reliable communication system.

Although some of the distributed intelligence based approaches were invented decades ago, they still have application value. But with increasing scale and interconnect degree, use of distributed intelligence based devices has become more and more difficult.

Fault Location and Service Restoration for Electrical Distribution Systems, First Edition. Jian Liu, Xinzhou Dong, Xingying Chen, Xiangqian Tong, Xiaoqing Zhang and Shiming Xu.

This chapter begins with the basic principle of FA based on recloser and voltage-delay type sectionalizers in Section 3.2. Section 3.3 then deals with the basic principle of FA based on reclosing with a fast over-current protection mode. Section 3.4 describes the fast healing approach based on neighbor communication.

3.2 FA Based on Recloser and Voltage-Delay Type Sectionalizers

The feeder automation system with coordinated reclosers and voltage-delay type sectionalizers is typical of distribution automation systems without a master station. It has the advantages of low costs, reliable action, and so on. This system is suitable for radial grids, "hand-in-hand" open-loop grids, and multi-sectioned and multi-linked grids. It is not suitable for more complicated grid structures. The appropriate parameter settings for reclosers and voltage-delay type sectionalizers are key for using this system and if the settings are inappropriate, the scope of fault isolation will be expanded and the time taken for service restoration will be extended.

The basic principle of coordination for reclosers and voltage-delay type sectionalizers will be introduced in this section. The approach to setting the parameters for them will not be discussed in this book, which can be found in [12].

In the coordinated feeder automation system, the two overlapping functional reclosers are used for the power switches of the feeders, the first delay time which is relatively longer (typically 15 sec.) and the second relatively shorter (typically 5 sec.). Voltage-delay type sectionalizers are used for sectional and interconnection switches (i.e., the loop switches).

The voltage-delay type sectionalizer is composed of a switch (SW) body, two single-phase-source (SPS) transformers, and a fault detecting relay (FDR), as shown in Fig. 3.1.

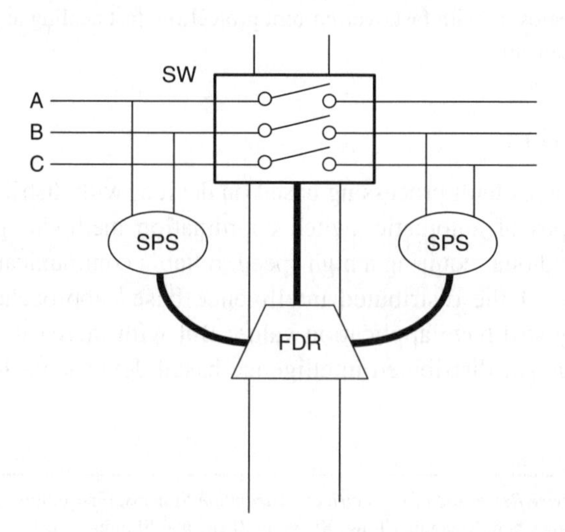

Figure 3.1 Composition of the voltage time type sectionalizer

The voltage-delay type sectionalizer generally has two sets of functions, one set of which is aimed at the sectional switches in the normally closed state, the other set applies to the interconnection switches; that is, the loop switches in the normally open state. The two sets of functions can be interchanged through a handle.

The voltage-delay sectionalizer should be set to the first set of functions when it is used for the sectional switch. The sectionalizer should be switched off when it detects loss of voltage. The X-counter will start when one side of the sectionalizer detects voltage fed to it. The sectionalizer should be turned on when a time delay of X is reached. The Y-counter should be started at the same time. If the sectionalizer is tripped due to loss of voltage within the period of Y time, it will be locked to be in the OFF state and will not reclose any more.

The voltage-delay type sectionalizer used as the interconnection switch should be set to the second set of functions. The X_L counter should be started when the Fault Detection Relay (FDR) detects a voltage loss on any side of the interconnection switch. The interconnection switch should be switched on when the time delay of X_L is reached. The Y_L counter should be started at the same time. If the interconnection switch is tripped due to loss of voltage within the period of Y_L, it will be locked to be in the OFF state and will not reclose anymore.

When a fault occurs, the recloser used as the power switch on the fault feeder will trip and then all of the sectionalizers on the fault feeder will switch off due to loss of voltage. Then, the recloser will reclose the first time after the delay of 15 sec. The sectional switches will switch on, respectively, when the voltage fed to them and the corresponding time delays are reached.

If it is a temporary fault, all of the sectionalizers on the fault feeder will be successfully reclosed and all of the service is restored.

In case of a permanent fault, the recloser will trip the second time when the sectionalizer connecting the fault region is turned on, then all of the sectionalizers on the fault feeder will switch off again due to loss of voltage. The sectionalizer connecting with the fault region will be locked to the OFF state because it does not stay in the ON state for more than the Y-time. The other sectionalizers connecting with the fault region are also locked to their OFF state due to the residual voltage locking mechanism. The recloser on the fault feeder will reclose the second time 5 sec. later and the corresponding sectionalizers will switch on in turn, but the sectionalizer connecting the fault region remains its OFF state. Thus, the service upstream of the fault region is restored. When the time delay of X_L is reached, the interconnection switch will be turned on and the sectionalizers downstream of the fault region will switch on in turn and the service of the regions downstream of the fault is restored.

The fault processing of a typical radial distribution grid is shown in Fig. 3.2. A is the recloser with its two reclosing delay times of 15 sec. and 5 sec. B, C, D, and E are the voltage-delay type sectionalizers, setting at the first set of functions. The X-times of B and D are both set to 7 sec. The X-times of C and E are both set to 14 sec. The Y-times of B, C, D, and E are all set to 5 sec.

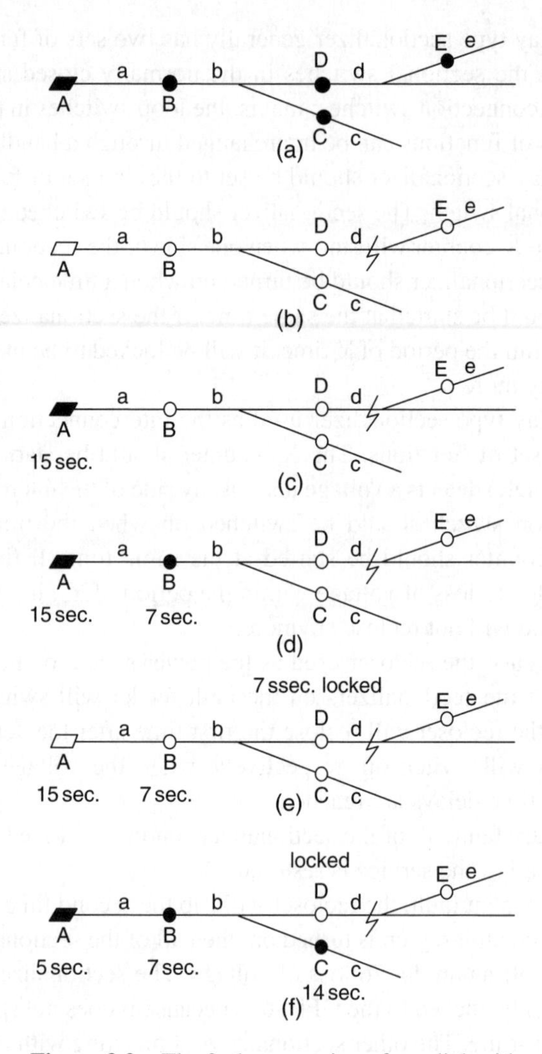

Figure 3.2 The fault processing of a radial grid

The normal situation of the radial grid is shown in Fig. 3.2(a). When a permanent fault occurs in region D, the recloser A trips making sectionalizers B, C , D, and E switch off (shown in Fig. 3.2(b)). After 15 sec., recloser A recloses the first time (shown in Fig. 3.2(c)). Sectionalizer B recloses 7 sec. later (shown in Fig. 3.2(d)). Then, 7 sec. later, sectionalizer D recloses and feeds the power to the fault region causing recloser A to be tripped again. Sectionalizer B and D trip due to loss of voltage and D is locked to the OFF state shown in Fig. 3.2(e). Recloser A recloses the second time 5 sec. later, sectionalizer B recloses 7 sec. later, and sectionalizer C recloses 14 sec. later. Fault processing is completed as shown in Fig. 3.2(f).

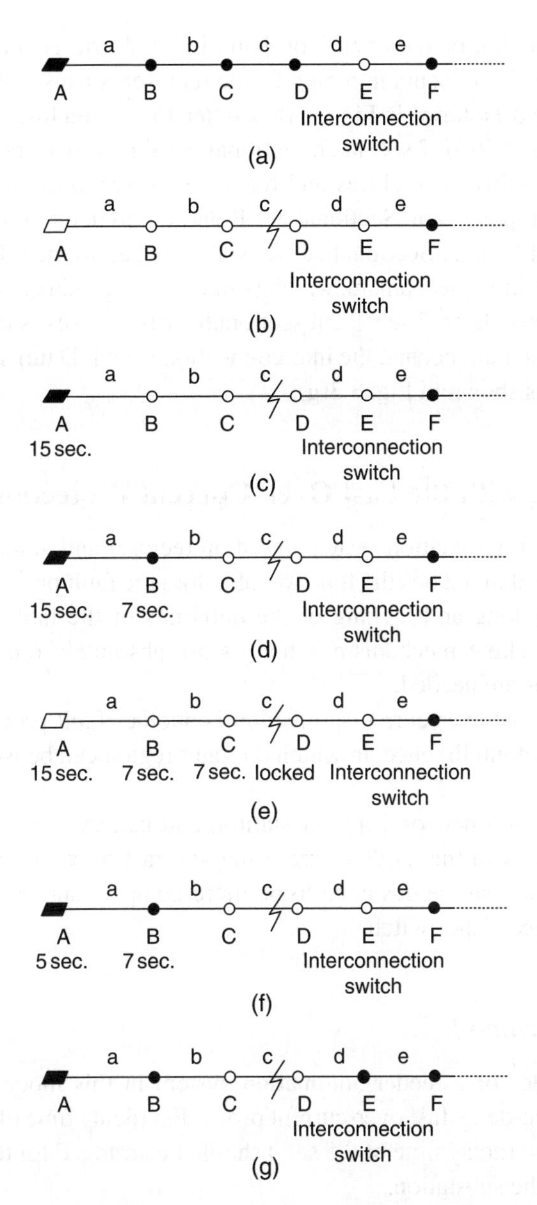

Figure 3.3 The fault processing of an open-loop hand-in-hand grid

The fault processing of an open-loop hand-in-hand distribution grid is shown in Fig. 3.3. A is the recloser with its two reclosing delay times of 15 sec. and 5 sec. B, C, D, and F are the voltage-delay type sectionalizers, set at the first set of functions with the X-time of 7 sec. E is the voltage-delay type sectionalizer set at the second set of functions with the X_L-time of 45 sec. The Y- and Y_L-time are both 5 sec.

The normal situation of the open-loop hand-in-hand grid is shown in Fig. 3.3(a). When a permanent fault occurs in region C, the recloser A trips making sectionalizers B, C, and D switch off (shown in Fig. 3.3(b)). After 15 sec., recloser A recloses the first time (shown in Fig. 3.3(c)): 7 sec. later, sectionalizer B recloses (shown in Fig. 3.3(d)); 7 sec. later, sectionalizer C recloses and feeds the power to the fault region causing recloser A to be tripped again. Sectionalizer B and C trip due to loss of voltage and C is locked to the OFF state. Sectionalizer D is also locked to the OFF state due to the residual voltage locking mechanism, which is shown in Fig. 3.3(e). Recloser A recloses the second time 5 sec. later, 7 sec. later, sectionalizer B recloses (shown in Fig. 3.3(f)), and 45 sec. after the fault occurs, the interconnection switch D turns on. Fault processing is completed as shown in Fig. 3.3(g).

3.3 Reclosing with the Fast Over-Current Protection Mode

Although the feeder automation system based on recloser and voltage-delay type sectionalizers described in 3.2 has the function of automatic fault processing, the processing time is a little long and locking of the outpoints of the fault region relies on a residual voltage locking mechanism, which is not absolutely reliable. Besides, two reclosing processes are needed.

Reclosing with fast over-current protection is another fault processing technology based on distributed intelligence, in which the fault region can be isolated by only one reclose process.

The communication network, master station, and battery are not used in a feeder automation system with the mode of reclosing and fast over-current protection, thus the switch controller can be designed to be more compact and can even be installed directly into the box of the switch.

3.3.1 Basic Principle

The basic principles of a feeder automation system in this mode are shown in the following. The time-delay fast over-current protection (delay time of 200~300 ms) and a reclosing function (delay time of 0.5 sec.) should be arranged for the 10 kV outgoing circuit breaker in the substation.

All of the sectional switches on the feeder should be circuit breakers set to have the first set of functions, such as automatic switch off when the loss of voltage being detected, reclose after a time-delay of t_I when one side is energized, instantaneous over-current protection to trip when detecting the fault current during the period of t_y (t_y is generally 1~2 sec.) after reclosing and the corresponding switch locked in its off state, instantaneous over-current protection being turned off after t_y, and switch-on prohibited when two sides of a switch in the OFF state are both energized.

The interconnection switch (i.e., the loop switch) should be a circuit breaker set to have the second set of functions, such as switching on after the time-delay of t_{II} in case

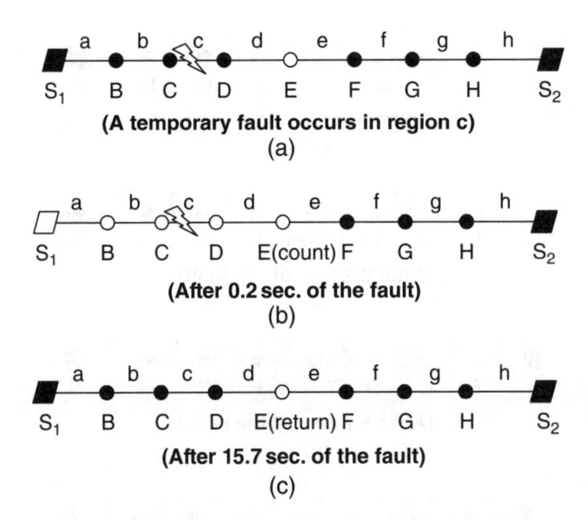

Figure 3.4 The temporary fault processing for reclosing with fast over-current protection mode

of a voltage loss on any side of the interconnection switch is detected, instantaneous over-current protection to trip when detecting the fault current during the period of t_y (t_y is generally 1~2 sec.) after reclosing and the corresponding switch locked in its off state, and switch-on is prohibited and returns to its original state when two sides of a switch in the OFF state are both energized.

As for the examples shown in Fig. 3.4(a) and 3.4(b), S_1 and S_2 denote the 10 kV outgoing switch in the substation; the delay time for the over-current protection is 0.2 sec.; B, H, C, D, F, and G are the sectional switches, respectively; and E is the interconnection switch. In these two figures, the solid symbols denote the switch in its ON state, the hollow symbols denote the switch in its OFF state.

B, H, C, D, F, and G should be set to have the first set of functions, where t_I is set to 5 sec. and t_y is set to 2 sec. E should be set to have the second set of functions, where t_{II} is set to a value of 20 sec. and t_y is set to a value of 2 sec.

If a temporary fault occurs in region c, S_1 will switch off due to over-current protection after a time-delay set beforehand. Then, B, C, and D will switch off due to the loss of voltage and the switch-on counter of E will start, as shown in Fig. 3.4(b). After 0.5 sec., S_1 recloses successfully and delivers power to B. 5 sec. later B will switch on and deliver power to C, 5 sec. later, C will switch on and deliver power to D, 5 sec. later, D will switch on successfully, and E returns to its original state, as shown in Fig. 3.4(c).

If a permanent fault occurs in region c, S_1 will switch off due to over-current protection after a time-delay set beforehand. Then, B, C, and D will switch off due to the loss of voltage and the switch on counter of E will start, as shown in Fig. 3.5(b). After 0.5 sec., S_1 will reclose successfully and deliver power to B. 5 sec. later, B will switch on and deliver power to C. The instantaneous over-current protection of B will be locked after 2 sec., as shown in Fig. 3.5(c). 5 sec. later, C will switch on and its instantaneous over-current

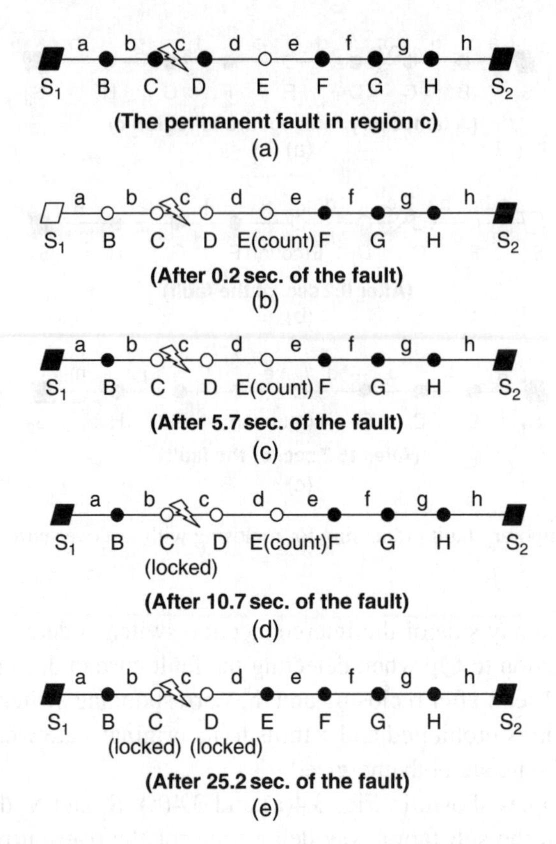

Figure 3.5 The permanent fault processing for reclosing with fast over-current protection mode

protection will trip the switch and C will be locked in its OFF state, as shown in Fig. 3.5(d). 20.2 sec. after the fault, E reaches its delayed time. Then E will switch on and deliver power to D. 5 sec. later, D will switch on and its instantaneous over-current protection will trip the switch and D will be locked in its OFF state. But the protection of S_2 will not start due to the time delay. The result of permanent fault processing is shown in Fig. 3.5(e).

3.3.2 *Improvements*

The basic approach of reclosing with the fast over-current protection mode has the following shortcomings:

- In permanent fault processing, after the interconnection switch is closed to restore downstream of the fault region by the other feeder, over-current protection will trip the corresponding switch once it is over-loaded.

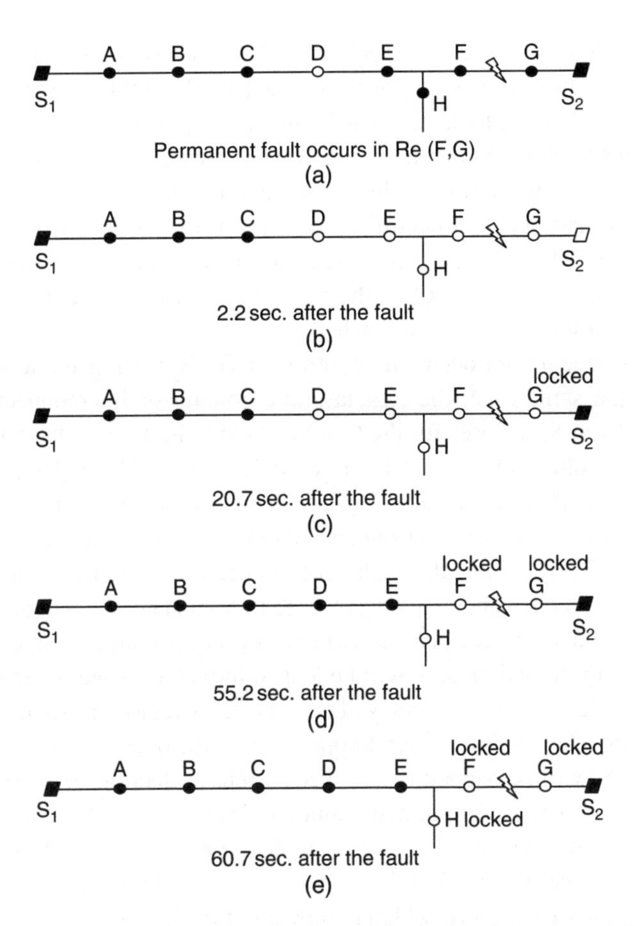

Figure 3.6 The permanent fault processing for the improved reclosing with fast over-current protection mode

- The duration time of temporary fault processing is still too long.
  To solve this problem, the following improvements can be made.
  - The recloser in the substation is improved to be allowed to reclose twice with the delay times of 0.5 sec. and 10–15 sec., respectively.
  - The automatic switch off function of the sectional switch is improved to have a delay time of 1 to 1–2 sec. when loss of voltage is detected.
- As for the sectional switches and the interconnection switches, Y-time lock and Y_L-time lock functions are added. If the two sides of a switch cannot maintain the energized state over the period of Y-time or Y_L-time, it should be tripped immediately and locked to be in the OFF state so it will not reclose anymore.

As for the example in Fig. 3.6(a), the delay times of the over-current protection of the outgoing switches of S_1 and S_2 are both 0.2 sec. The delay times of the twice-reclosing of S_1 and S_2 are 0.5 and 15 sec., respectively. All of the sectional switches of A, B, C, E, F, and G have setting values of t_I of 5 sec. The t_I of H is set to 10 sec. The t_{II} of interconnection switch D is set to 45 sec. The Y-time and Y_L-time are all set to 3 sec. The t_y of the sectional switches and the interconnection switch are set to 2 sec. The automatic switch-off delay time when there is loss of voltage is set to 1.5 sec.

In case of a temporary fault occurring in region Re(F,G), S_2 is tripped due to over-current protection. 0.5 sec. later, S_2 is reclosed to restore the service of the whole feeder, since G, F, E, and H are still in their ON state, because the delay time of 1.5 sec. (tripping due to voltage loss) is not reached.

In case a permanent fault occurs in region Re(F,G), S_2 is tripped due to over-current protection; at the same time, the t_{II} delay time counter of interconnection switch D starts. 0.5 sec. later, S_2 recloses for the first time and fails. 1.5 sec. later, G, F, E, and H switch off due to voltage loss, which is shown in Fig. 3.6(b). 15 sec. later, S_2 recloses for the second time and delivers power to G, 5 sec. later G switches on and its instantaneous over-current protection will trip the switch and G will be locked in its OFF state (shown in Fig. 3.6(c)), 45.2 sec. after the fault, interconnection switch D switches on as the delay time of t_{II} is reached and delivers power to E. 5 sec. later, E switches on and delivers power to F and H. 5 sec. later, F switches on and its instantaneous over-current protection will trip the switch so F will be locked in its OFF state but the protection of S_1 will not start due to the time delay. Thus, the fault region is isolated as shown in Fig. 3.6(d). H switches on 5 sec. later. Supposing that the overload of S_1 is caused when H switches on, S_1 will be tripped by its over-current protection. H is tripped immediately due to the two sides of it not maintaining the energized state over the Y period of time, which is set to 5 sec., and it is locked in the OFF state. But A, B, C, D, and E are not tripped. S_1 recloses for the first time 0.5 sec. later to restore service but without the region downstream of switch H, which is shown in Fig. 3.6(e).

As for the basic approach without the improvement of the Y- and Y_L-time locks, S_1 would be reclosed and tripped repeatedly until the allowed reclosing times are used up. As a result, the service of the whole feeder from S_1 will not be restored.

3.4 Fast Healing Approach Based on Neighbor Communication

With the development of intelligent substation technology, high-speed network communications based on GOOSE have gradually matured. The fast healing approach based on neighbor communication may achieve rapid removal of fault and does not make the non-faulty area blackout.

3.4.1 Basic Principle

The over-current protections are configured to the outgoing switches in the substation. The Intelligent Electronic Devices (IED) (with functions described later in this section) are configured to the circuit breakers on the feeder. A IED of a feeder switch

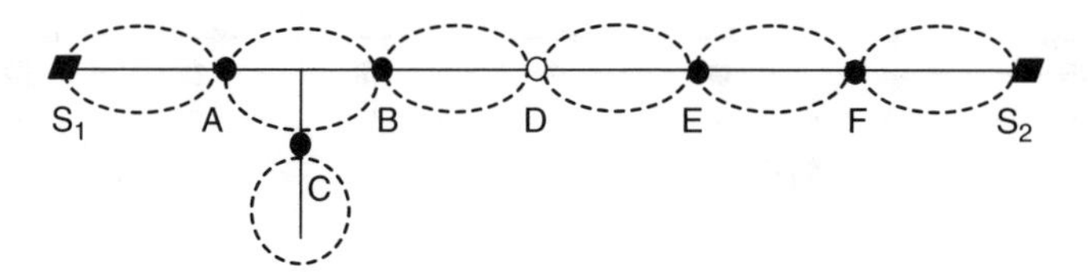

Figure 3.7 A typical open-loop distribution network

may send the fault information to the IEDs of its adjacent switches and identify whether the feeder switch should be tripped. Communication systems with high speed and reliability, such as, GOOSE, is used for communication among adjacent IEDs.

3.4.1.1 The Self-Healing Control of an Open-Loop Distribution Network

3.4.1.1.1 The Mechanism of Fault Location

If one phase of a switch flowing through the fault current is over the set value, its IED sends the fault information of "over-current" to the IEDs installed on the adjacent switches.

When only the enter point of a region has fault current flowing through, while none of the outpoints of the region have fault currents flowing through, the fault occurs in the region. The details of this rule are described in Chapter 4.

For the example shown in Fig. 3.7, if only the IED on A reports the fault current while the IEDs on B and C do not, the fault is in the region of Re(A, B, C). If both the IEDs on A and C report the fault current while the IED on B does not, the fault is not in the region of Re(A, B, C).

3.4.1.1.2 Mechanism of Fault Isolation

For an open loop distribution grid, the fault isolation mechanism is as follows:

- If a switch is one terminal of the fault region, it should be tripped.
- If none of the regions with the switch as terminal is the fault region, the switch should not be tripped even if fault current flowing through it.

Since the communication among adjacent IEDs is through the optical self-healing ring network based on GOOSE, fault information may be collected to make a decision within a short period of time, such as 15 ms. Then, the fault region is isolated by tripping all of its terminal switches.

3.4.1.1.3 Mechanism of Service Restoration

An interconnection switch, that is, the loop switch, usually stays in the open state with both sides energized. When a fault occurs on the feeder connected to one side of an interconnection switch, the voltage on the corresponding side of the interconnection

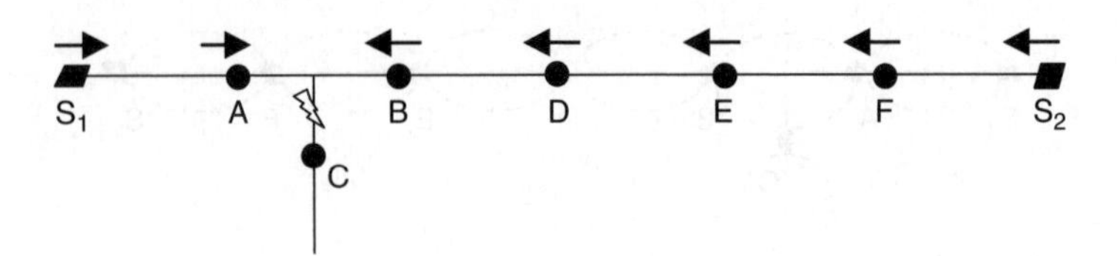

Figure 3.8 A typical close-loop distribution network

switch drops to zero, then the service restoration program starts to automatically close the interconnection switch after a time delay set beforehand (typically 5–15 sec.). Therefore, the service of the healthy regions downstream of the fault region may be restored.

3.4.1.2 The Self-Healing Control of a Close-Loop Distribution Network

3.4.1.2.1 Mechanism of Fault Location

If one phase of a switch flowing through the fault current over the set value, its IED sends the fault information to the IEDs installed on the adjacent switches in which the direction of the fault power flow is reported.

Where at least one of the terminals of a region have fault current flowing through along the direction of the fault power flow from the outside to inside of the region, while none of the terminals of the region with fault current flowing through along the direction of the fault power flow from the inside to the outside of the region, the fault occurs in the region. The details of this rule are described in Chapter 4.

For the example shown in Fig. 3.8, if the IEDs on A and B both report the direction of fault power flow from the outside of the region Re(A, B, C) to inside of the region, the fault is in the region of Re(A, B, C). Since no fault current is flowing through C, it dose not report any fault information. But, if the IED on A reports the direction of fault power flow from the outside of the region Re(A, B, C) to inside of the region, while the IED on B reports the direction of fault power flow from the inside of the region Re(A, B, C) to outside of the region, the fault is not in the region of Re(A, B, C).

3.4.1.2.2 Mechanism of Fault Isolation

The mechanism of fault isolation for close-loop distribution grids is the same as that for open-loop distribution grids.

3.4.1.2.3 Mechanism of Service Restoration

As for the close-loop distribution grids, once the fault region is isolated, the service to all of the healthy regions is restored, except for healthy regions on the radial branch downstream of the fault region.

As for the example shown in Fig. 3.9, two cascade power distribution switching stations are in the dotted boxes. S_1 and S_2 are the outgoing switches in the substations.

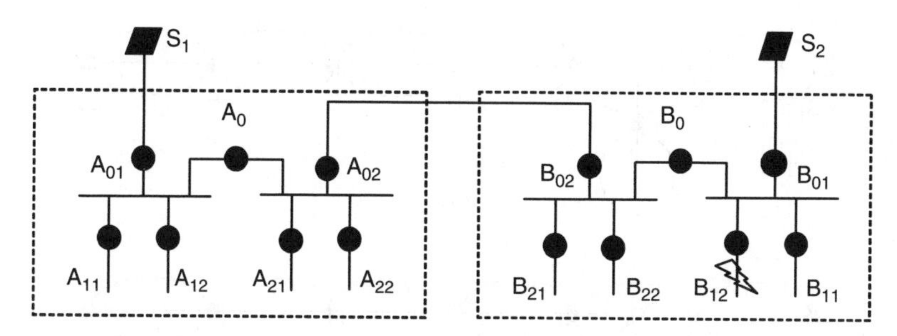

Figure 3.9 Two cascade power distribution switching stations

A_{01}, A_{02}, and B_{01}, B_{02} are the input switches of the two switching stations. A_0 and B_0 are the bus interconnection switches. A_{11}–A_{22} and B_{11}–B_{22} are the outgoing switches of the two switching stations. All of the other switches are in the on-state.

The IEDs with the fast healing approach based on neighbour communication technology are installed in each of the switching stations, respectively.

When a fault in the region Re(B_{12},-) occurs, the IED on B_{12} may identify that the fault is in the region with B_{12} as its terminal. Thus, B_{12} is tripped rapidly. Although the IEDs on B_0, B_{02}, B_{01}, S_2, A_{02}, A_0, A_{01}, and S_1 may also detect the fault current flowing through the corresponding switches, the IEDs identify that none of the corresponding switches is the terminal of the fault region, so they should not be tripped. Thus, over-tripping can be avoided.

3.4.2 Improvements

3.4.2.1 Improvements on Temporary Fault Processing

In case of a temporary fault, the basic fast-healing approach based on neighbor communication can also isolate the fault region causing unnecessary outage. To improve the performance of temporary fault processing, the following improvements are needed.

- The tripped circuit breaker with fault current is arranged to have a one-shot reclosing function when one side of it energized. If the reclosing fails causing the corresponding circuit breaker to be tripped again, it is locked in the OFF state and a message of *reclosing failure* is sent to its adjacent circuit breakers. If the reclosing is successful, a message of *reclosing success* is sent to its adjacent circuit breakers and the corresponding circuit breaker closes the function of self-healing control within 3 sec.
- Although tripped, the circuit breakers without the fault current do not have the reclosing function. Within a short period of 2 s after tripping, if it receives the message of *reclosing failure* or does not receive any message, it should stay in its

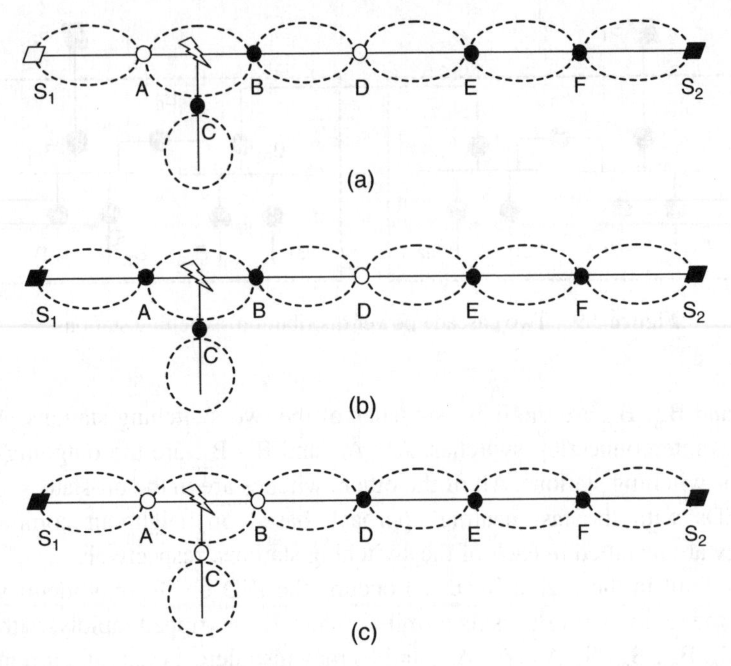

Figure 3.10 Fault processing in case of some of the fault information lost

OFF state. If it receives the message of *reclosing success*, it should close to its ON state.

- Un-tripped circuit breakers and successfully reclosed circuit breakers should ignore the *reclosing success message*.

Above improvements may also correct the mistake due to lost of fault information. See the example shown in Fig. 3.10(a), assuming that a fault occurs in the region Re(A,B,C). Some distractions cause the circuit breaker A to fail to send the fault information. According to Section 3.4.1, the region Re(S_1,A) is identified as the fault region causing S_1 and A tripping, as shown in Fig. 3.10(a). S_1 is reclosed successfully 0.5 sec. later and sends the message of *reclosing success* to A. Thus, A is closed, as shown in Fig. 3.10(b). If it is a temporary fault, the reclosing of A is successful and the service to the whole feeder is restored. In case of a permanent fault, the reclosing of A fails. The circuit breakers A, B, and C will trip to isolate the fault region of Re(A,B,C). The service of the region Re(S_1,A) is maintained. The interconnection switch D closes to restore the service of region Re (B,D) 10 sec. after the fault, as shown in Fig. 3.10(c).

3.4.2.2 Coping with Failure to Operate Circuit Breakers

Circuit breaker failure may destroy the fault processing of the basic fast-healing approach based on neighbor communication, to cope with this, the following improvements are needed.

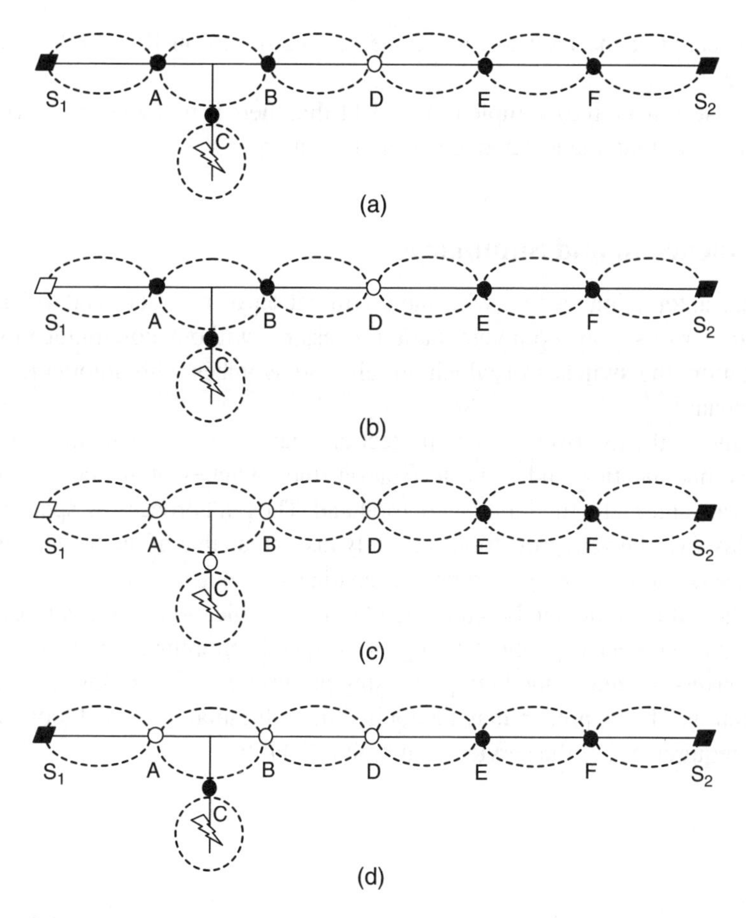

Figure 3.11 Permanent fault processing in case of a circuit breaker failed to operate

- If a circuit breaker is not closed or tripped after the corresponding control message is sent to it more than 200 ms, this is regarded as failure to operate.
- If a circuit breaker fails to operate, it should send a message of *fail to operate* to its adjacent circuit breakers.
- If a circuit breaker is in its ON state, it should trip once it receives the message of *fail to operate* from its adjacent circuit breaker.

See the example shown in Fig. 3.11(a), assuming that a permanent fault occurs in region Re(C,-) and circuit breaker C fails to operate. The over-current protection with time delay of the outgoing circuit breaker S_1 is tripped to clear the fault, as shown in Fig. 3.11(c). The IED on circuit breaker C sends the *fail to operate* message to its adjacent circuit breakers A and B. Circuit breakers A and B receive the message and trip, as shown in Fig. 3.11(c). S_1 is reclosed successfully 0.5 sec. later to restore the service of region Re(S_1,A), as shown in Fig. 3.11(d). The interconnection switch D closes 10 sec. after the fault to restore the service of region Re (B,D), as shown in Fig. 3.11(e).

It can be seen from the example in Fig. 3.11 that there is no more influence on fault processing other than the isolated area is a little enlarged.

3.5 Conclusion and Summary

The feeder automation system with the coordination of reclosers and voltage-delay type sectionalizers may complete fault processing without communication and a master station, the switches of which are all load switches with automatic functions set beforehand.

Reclosing with fast over-current protection may also complete fault processing without communication and a master station, the switches of which are all circuit breakers with automatic functions set beforehand. The fault processing speed of reclosing with fast over-current protection is usually faster than that of the feeder automation with reclosers and voltage-delay type sectionalizers

The fast-healing approach based on neigbor communication may clear out the fault rapidly and avoid over-tripping. Although high speed communication among adjacent circuit breakers is needed, the fault processing program may be completed without the master station. This approach may be applied in both radial grids and looped grids to meet the requirement of the service with high reliability.

4

Fault Processing Based on Centralized Intelligence

Liu Jian and Chen Xingying

Abstract
This chapter discusses fault processing based on a Distribution Automation System (DAS) with centralized intelligence, including fault location and service restoration. A simplified model for fault location and service restoration is established. The criteria for interphase short circuit fault location and isolation are described. The methodologies of service restoration and control sequence generation are also dealt with.

Keywords
Distribution Automation System (DAS), centralized intelligence, interphase short circuit fault, fault location, fault isolation, service restoration, Distribution Generation (DG), control sequence generation, simplified model, minimum load shedding, correction control in cases of failure of remote control

4.1 Introduction

This chapter discusses fault processing based on a Distribution Automation System (DAS) with centralized intelligence, including fault location and service restoration.

Although the relay protection and automatic switch coordination approaches have the advantages of rapid response and high reliability, fault location cannot be accurately realized for a feeder with numerous sections and, also, service restoration cannot

Fault Location and Service Restoration for Electrical Distribution Systems, First Edition. Jian Liu, Xinzhou Dong, Xingying Chen, Xiangqian Tong, Xiaoqing Zhang and Shiming Xu.

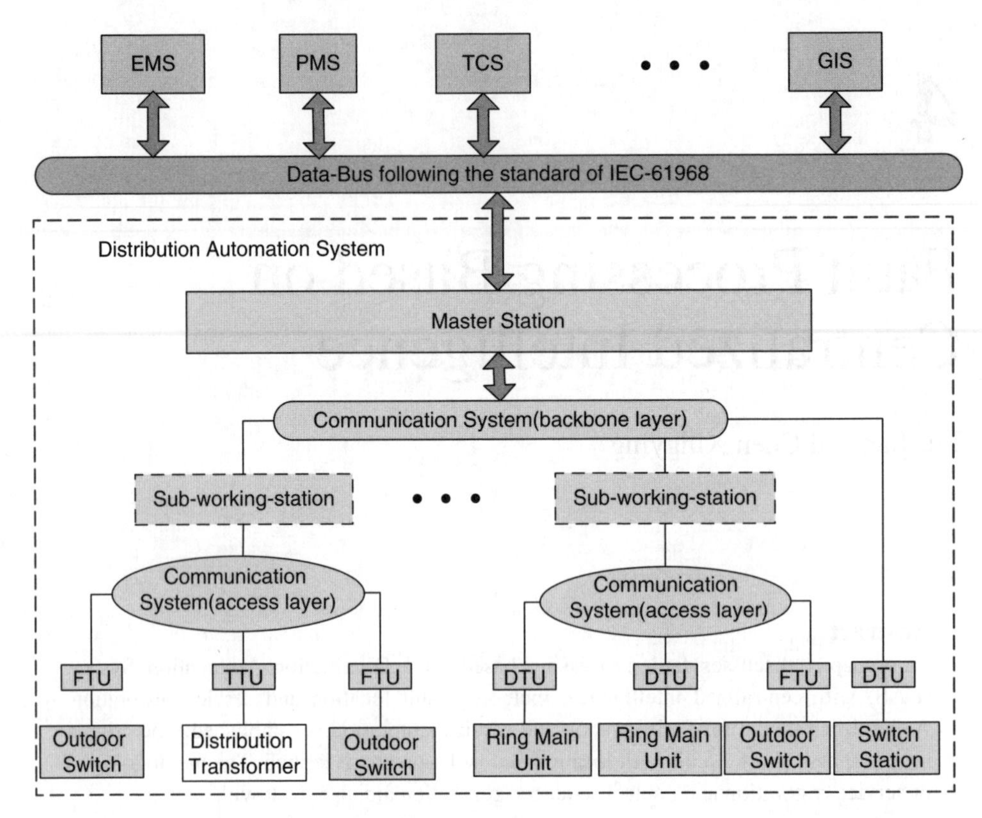

Figure 4.1 A DAS and its information interchanging with the corresponding systems

be optimized for complicated distribution grids and some specially designed grids, such as three-sectioned and three-linked grids, and three-supplying and one-back-up grids.

A *centralized intelligence based Distribution Automation System* (DAS) is generally composed of a *master station*, some *sub-stations*, a large number of *Feeder Terminal units* (FTU), and the communication system connecting the data transmission within the systems, which is shown in Figure 4.1 in the dashed block.

The master station is usually based on a computer network with servers, work stations, communication processors, and so on. It is the information and control center of the DAS with centralized intelligence based on global information from the whole distribution grid.

The sub-working-stations are important and have the function of concentrating information from the FTUs in a certain scope. Some sub-working-stations also play the roles of sub-master stations, which have the function of fault processing control and operation optimizing control of distribution grids with corresponding scopes.

Terminal units are outdoor devices that are usually mounted on the poles or in the ring main unit cabinets. Terminal units can be classified into controllable terminal units, such as Feeder Terminal Units (FTU), and un-controllable terminal units, such as fault indicators. With controllable terminal units, the master station of a DAS cannot only collect the information about voltages, currents, switch states, and faults of distribution grids, but also remotely control the corresponding switches. An uncontrollable terminal unit does not have the remote control function. The essential function of an uncontrollable terminal unit is sending fault information to the master station, although some uncontrollable terminal units may also have the functions of remote information collection about voltage, current, and switch state.

Once a fault occurs, a certain circuit breaker is quickly tripped by the relay protection device to interrupt the circuit. The *fault processing program* starts when the master station receives information from a certain circuit breaker tripped while its corresponding relay protection device actuates. In other words, a particular circuit breaker tripped by the corresponding relay protection device is the *starting condition of a fault processing program for a DAS.*

The DAS fault processing program is composed of the following three processes in turn.

1. *Fault information collecting process*, in which the master station spends a certain time, typically 30 sec. set in advance, collecting fault information from the corresponding terminal units and from the sub-station automation system.
2. *Fault location process*, in which the master station locates the fault position by centralized intelligence according to fault information.
3. *Fault isolation and service restoration process*, in which the master station works out the optimal fault isolation and service restoration scheme, and executes the scheme by remote control.

Typically, a DAS is linked to other information systems, such as the Geo-Information System (GIS), Energy Management System (EMS), Production Management System (PMS), Trouble Call System (TCS), and so on. It is recommended that information exchanging should be based on a data-bus following the IEC-61968 standard.

A DAS and its information exchanging with the corresponding systems is shown in Figure 4.1.

The bulk of this chapter begins with a simplified model of distribution grids for fault location and service restoration in Section 4.2. Section 4.3 then deals with the criteria of interphase short circuit fault location and the approach to improving the tolerance of fault location in cases where some fault information is missing and/or wrong. The adaptabilities of fault location criteria in the case of Distributed Generation (DG) access are analyzed. Fault location methodologies meeting the challenge of large scale of Distributed Generation (DG) access are also discussed. Section 4.4 begins with a discussion on fault isolation. Then four types of service restoration, such as real time service restoration, short time service restoration, service restoration with minimum load shedding, and modeled restoration, are detailed together with their relationships. The methodologies of control sequence generation and correction control in cases of remote control failure are also dealt with.

4.2 Simplified Modeling of Distribution Grids

A simplified model of distribution grid for fault location and service restoration is established in this section. The number of nodes is remarkably reduced but load flow analysis is approximated. Nonetheless, the simplified model is sufficient for fault location since all information needed is included with the advantage of clarity. It is also suitable for service restoration because the approximated evaluation of restoration programs is enough and the model is also fast.

The knowledge described in this section is the foundation for the following sections.

4.2.1 Distribution Network Structure

The simplified model is based on graph theory. A distribution grid is regarded as a weighted graph. The switches and the points with fault monitoring devices are regarded as nodes. The distribution lines connecting the adjacent nodes are regarded as edges.

The load flowing through a node is defined as the weight of the corresponding node. The load distributed on an edge is defined as the weight of the corresponding edge.

In this section, the concepts of the simplified distribution network structure modeling are described but the concrete data structures of the model are not dealt with since there are so many approaches to establishing data structures. References [6] and [11] describe some feasible data structures, such as the adjacency matrix approach and the adjacency list approach, to name a couple.

4.2.1.1 Nodes

Nodes can be classified into four types as follows:

1. *Action nodes* are the nodes of controllable switches, which can be controlled by the corresponding Feeder Terminal Units (FTU) to turn on or off the corresponding switches to connect or break the circuits. Action nodes can be classified into two kinds; sectionalizing action nodes based on the sectionalizing switches that are normally closed and loop action nodes based on the loop switches that are normally open.
2. *Monitoring nodes* are nodes where fault identification devices are installed. Usually, an action node is also a monitoring node. But sometimes a monitoring node is not an action node; that is, a fault identification device is installed on a position without any switch, such as the beginning of a branch or a lateral, the medium of a rather long branch, and so on.

Table 4.1 The relationship between the four types of node

	Switching	Monitoring	Terminal	Source
Action nodes	Always	Quite often	In case of cables	In substations
Monitoring nodes	Not necessary	Always	A few	In substations
Ending nodes	Never	Never	Always	Never
Source nodes	Always	Always	A few	Always

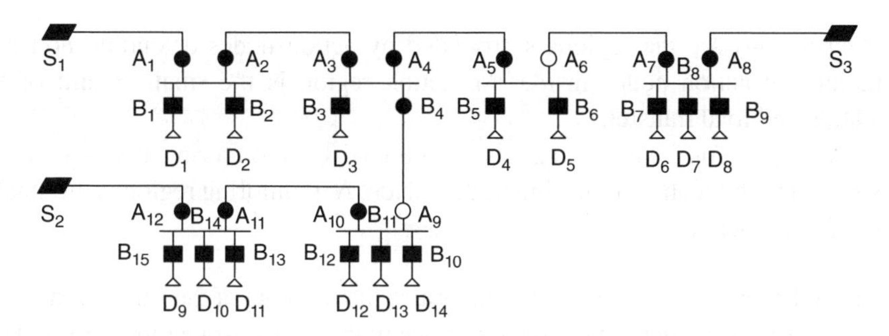

Figure 4.2 An example to illustrate the four types of nodes

3. *Ending nodes* are the terminal of a trunk, a branch, or a lateral. The state of an ending node is constantly open.
4. *Source nodes* are the nodes identifying the power source. Since an electric power distribution system can be regarded as an infinite power system, source nodes are often the action nodes of distribution circuit breakers in the substations. The switches connecting the Distribution Generations (DG) to the main distribution grid or PCC can also be regarded as source nodes. Thus, source nodes can be classified into two kinds; the *main source nodes* and the *DG source nodes*.

The relationship between the four types of node is shown in Table 4.1.

If the *i*-th node is directly connected with the *j*-th node; that is, there is no node on the route between the *i*-th node and the *j*-th node, we call that the *i*-th node and the *j*-th node are adjacent; that is, the *i*-th node is the adjacent node of the *j*-th node and vice versa.

In the example in Fig. 4.2, the solid icons indicate the corresponding switches are closed, while the hollow icons indicate the corresponding switches are open. In Fig. 4.2, the parallelograms indicate the source nodes that are outgoing circuit breakers in the substations; that is, they are the main source nodes. The circles indicate the action nodes that are sectionalizing switches with FTU installed along the trunks. Solid circles indicate the sectionalizing action nodes and hollow circles indicate the loop action nodes. With FTUs, the circles are the monitoring nodes as well. The squares indicate the monitoring nodes with fault indicators installed. The triangles are ending nodes. The adjacent nodes to node-B_3 are node-A_3, node-A_4, and node-B_4.

4.2.1.2 Regions

A region is a sub-graph surrounded by a certain type of nodes and has no nodes of the same type inside.

Because regions have clear physical meaning but edges haven't, the concept of region is generally used in the simplified model instead of the concept of edge. A region contains many edges inside.

Regions can be classified into two types as follows:

1. *Action regions* are the regions surrounded by action nodes or ending nodes and without any action nodes inside. An action region is the smallest unit of fault isolation and load transfer.
2. *Monitoring regions* are the regions surrounded by monitoring nodes or ending nodes and without any monitoring nodes inside. A monitoring region is the smallest unit of fault location.

If one of the terminals of the *i*-th region is also the terminal of the *j*-th region, we call that the *i*-th region and the *j*-th region are adjacent; that is, the *i*-th region is the adjacent region of the *j*-th region and vice versa.

In the example in Fig. 4.2, the action regions are shown in Fig. 4.3, which are the smallest units of fault isolation and load transfer. The adjacent regions of the region Re(A_3, A_4, D_3, B_4) are Re(A_2, A_3), Re(A_4, A_5), and Re(B_4, B_9). The monitoring regions are shown in Fig. 4.4, which are the smallest units of fault location. The adjacent regions of the region Re(A_3, A_4, B_3, B_4) are Re(A_2, A_3), Re(A_4, A_5), Re(B_3, D_3), and Re(B_4, B_9).

4.2.1.3 Feeders

A feeder is a sub-graph starting from a main source node and terminates at some ending nodes and loop switches, which are usually action nodes or monitoring nodes. This definition of a feeder is the same as the term usually used in practice.

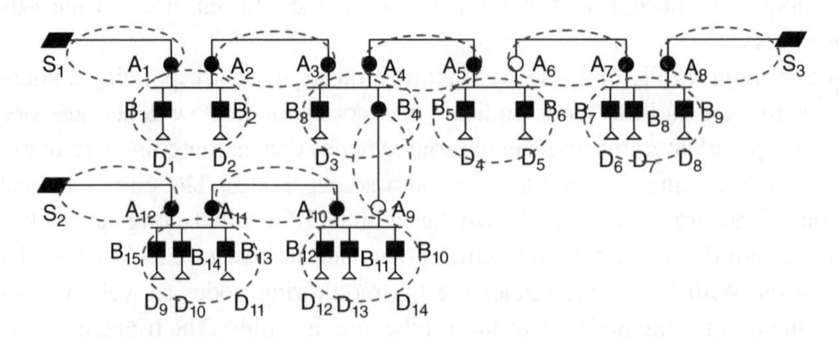

Figure 4.3 The action regions of the example in Fig. 4.2

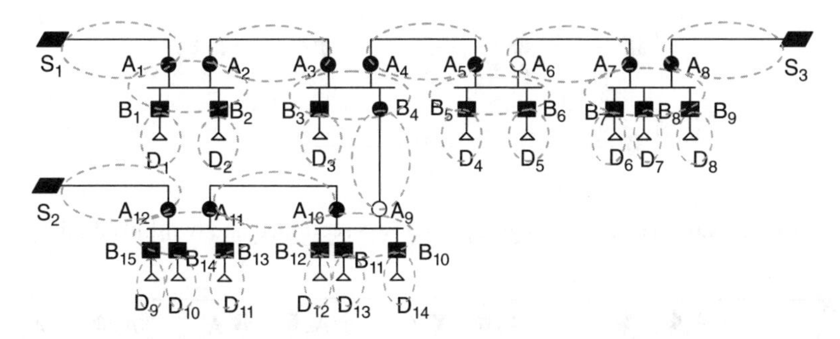

Figure 4.4 The monitoring regions of the example in Fig. 4.2

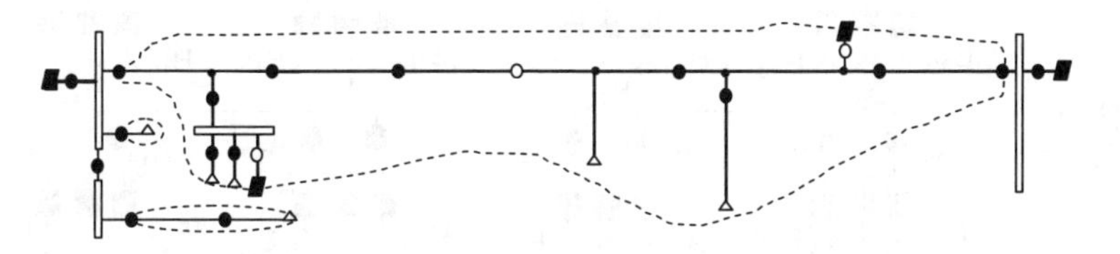

Figure 4.5 Dividing connected systems

4.2.1.4 Connected Systems

A connected system is a sub-graph with some feeders connected by loop-switches. A connected system has the following properties:

- The regions are all connected in a connected system when all of the loop switches are closed.
- The number of connected regions cannot be increased when any switch outside the corresponding connected system is closed.
- The connection through the buses in the substations are not considered; that is, the boundary nodes of a connected system may merely be the source nodes and the ending nodes.

A connected system is the smallest unit for network reconfiguration and service restoration. In other words, no matter how large scale a distribution grid is, its service restoration in cases of feeder faults may be carried out within a certain connected system, which is generally on a much smaller scale.

In the example in Fig. 4.5, the connected systems are illustrated by dashed rings. In the figure, the triangles indicate the ending nodes, the thin rectangles indicate the buses in the substations.

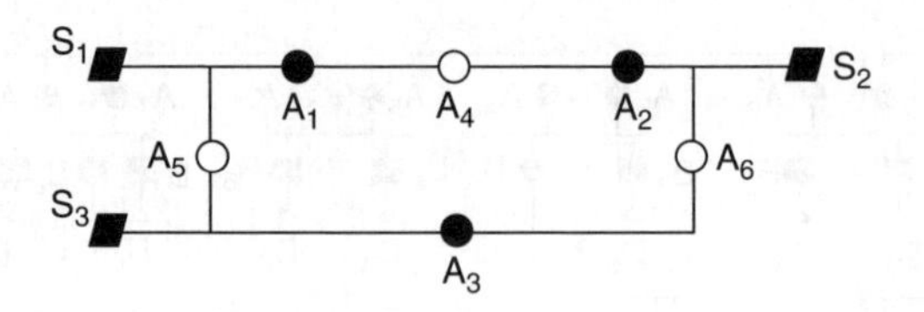

Figure 4.6 The modeled groups of a two-sectioned and two-linked grid

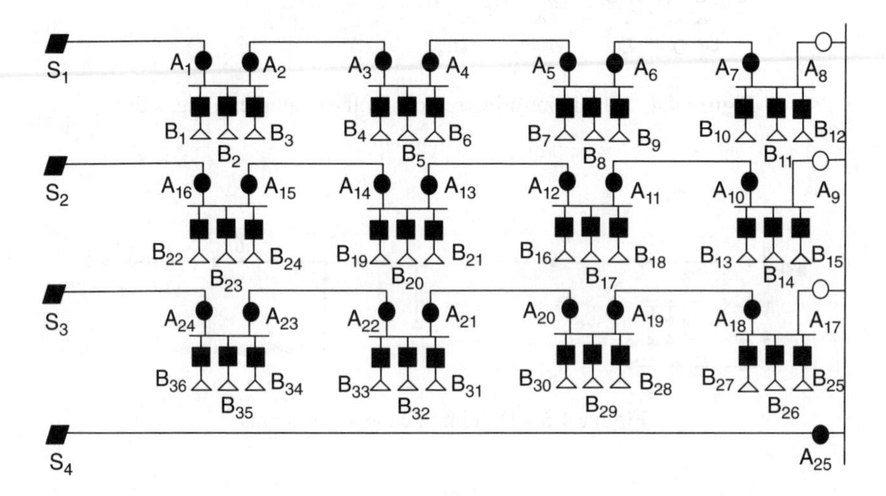

Figure 4.7 The modeled groups of a three supplying and one back-up grid

4.2.1.5 Modeled Groups

Sometimes, certain intentionally designed connections are built, such as three-sectioned and three-linked grids, three-supplying and one-back-up grids, and 4 × 6 connection grids to improve power supply capacity. In these cases, it is necessary to mark the related sub-graph so as to make fault isolation and service restoration convenient for the DAS. A modeled group is the sub-graph containing a certain intentionally designed connection.

The modeled groups of a two-sectioned and two-linked grid and a three-supplying and one-back-up grid are shown in Figs 4.6 and 4.7, respectively.

A special type of modeled group is the *conscious island* with some Distributed Generations (DG) and loads inside. This type of modeled group makes fault isolation and service restoration of distribution grids with DGs more convenient for the DAS.

4.2.1.6 Reference Direction

Due to the access of Distribution Generations (DG), the direction of the power flowing through a distribution grid appears bidirectional. Nonetheless, we define the direction from the main power source to the terminals of a feeder as the reference direction.

Thus, the terms *upstream* and *downstream* have meanings for single main source open loop distribution grids.

4.2.1.7 Node Hierarchies

Node hierarchies are defined for single main source open loop distribution grids.

The hierarchy of the *i-th* node, that is, $H_N(i)$ is defined as in Equation (4.1).

$$H_N(i) = \text{num\_N} + 1 \tag{4.1}$$

Where, num_N is the number of node from the i-th node to the main source node of the feeder.

The hierarchy of the main source node of the feeder is zero.

As for the i-th node and the j-th node, if

$$H_N(j) - H_N(i) = 0 \tag{4.2}$$

We know that the i-th node and the j-th node are *brothers*.

if

$$H_N(j) - H_N(i) = 1 \tag{4.3}$$

We call the i-th node the *father node* of the j-th node and the j-th node the *son node* of the i-th node.

if

$$H_N(j) - H_N(i) = 2 \tag{4.4}$$

We call the i-th node the *Grandfather node* of the j-th node and the j-th node the *Grandson node* of the i-th node.

if

$$H_N(j) - H_N(i) > 0 \tag{4.5}$$

We call the i-th node the *Elder node* of the j-th node and the j-th node the *Junior node* of the i-th node.

The *distance* between the i-th node and the j-th node, that is, $D_N(i,j)$ is defined as

$$D_N(i,j) = H_N(j) - H_N(i) \tag{4.6}$$

In the example shown in Fig. 4.6, the numbers in brackets indicate the hierarchies of the nodes, the numbers beside the nodes are the sequence numbers of the nodes, the dashed lines show the hierarchy dividing, and the arrows show the reference direction.

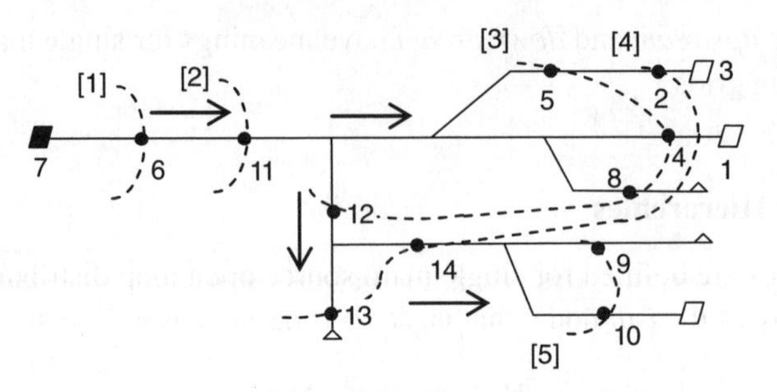

Figure 4.8 An example of node and region hierarchies

It can be seen from Fig. 4.8 that node-11 is the father node of node-4, node-5, node-8, and node-12.

4.2.1.8 Region Hierarchies

Region hierarchies are defined for single main source open loop distribution grids.

The nodes surrounding a region are called the *terminals of the region.* The terminal with the smallest hierarchy value is called the *enter point* of the region. The other terminals are called the *out points* of the region. A region can be expressed as Re (enter-point, out point-1, out point-2,…).

The hierarchy of the *i*-th region, that is, $H_R(i)$ is defined as the hierarchy value of its enter point.

As for the *i*-th region and the *j*-th region, if

$$H_R(j) - H_R(i) = 1 \tag{4.7}$$

We call the *i*-th region the *father region* of the *j*-th region and the *j*-th region the *son region* of the *i*-th region.

The definitions of brother, grandfather, grandson, elder, and junior regions, and the distances between the two regions are similar, as are the corresponding terms for nodes.

In the example shown in Figure 4.8, Re (11,4,5,8,12) is the father region of Re (12,13,14) and also is the grandfather region of Re (14,9,10).

4.2.2 *Simplified Load Flow Analysis*

In this section, a simplified approach to load flow analysis for a distribution grid is described, which is approximate but quite simple and fast so is suitable for service restoration program evaluation.

4.2.2.1 Load Descriptions

Strictly speaking, the *load flowing through a node* or *load distributed from a region* should be described as the form of complex power, that is, the combination of active power and reactive power.

The load through the *i*-th node, that is, $l_N(i)$, is

$$l_N(i) = s_N(i) = P_N(i) + jQ_N(i) \tag{4.8}$$

Where, $s_N(i)$, $P_N(i)$, and $Q_N(i)$ are the complex power, the active power, and the reactive power flowing through the *i*-th node, respectively. Of course, $P_N(i)$ and $Q_N(i)$ may be positive or negative, which means consuming or producing power and inductive reactive power or capacitive reactive power. As an example, if the active power through a node is positive, it means that downstream of the node consumes the corresponding amount of active power. If the active power through a node is negative, it means that downstream of the node produces the corresponding amount of active power.

The amplitude of $l_N(i)$, that is, $L_N(i)$ is

$$L_N(i) = |s_N(i)| = \sqrt{P_N^2(i) + Q_N^2(i)} \tag{4.9}$$

The load distributed from the *i*-th region, that is, $l_R(i)$, is

$$l_R(i) = s_R(i) = P_R(i) + jQ_R(i) \tag{4.10}$$

Where, $s_R(i)$, $P_R(i)$, and $Q_R(i)$ are the complex power, the active power, and the reactive power distributed from the *i*-th region, respectively. Of course, $P_R(i)$ and $Q_R(i)$ may also be positive or negative meaning the same as the load through a node.

The amplitude of $s_R(i)$, that is, $L_R(i)$ is

$$L_R(i) = |s_R(i)| = \sqrt{P_R^2(i) + Q_R^2(i)} \tag{4.11}$$

The capacity limitation of a node, usually being a switch, is generally described as the form of the current limitation (A). Thus, it is needed to convert the load flowing through a node into the current. As for the *i*-th node in the three-phase distribution grid, we have the load current through it, that is, $I_N(i)$ as

$$I_N(i) = \frac{S_N(i)}{\sqrt{3}U} = \frac{\sqrt{P_N^2(i) + Q_N^2(i)}}{\sqrt{3}U} \tag{4.12}$$

Where, U is the rated voltage of the distribution grid.

The capacity limitation of a region, usually being the summation of all of distribution transformers in the region, is generally described as the form of the apparent power limitation (kVA), thus, the amplitude of s_R, that is, S_R may do the work.

Sometimes, the relative loadings compared with capacity limitation are of importance, such as the cases of load balancing in the service restoration program.

Therefore, we define *relative loading* through the *i*-th node, that is, $I_N'(\mathrm{i})$ as

$$I_N'(\mathrm{i}) = \frac{I_N(\mathrm{i})}{I_{N,\max}(\mathrm{i})} \tag{4.13}$$

Where, $I_{\mathrm{N,max}}(\mathrm{i})$ is the current limitation of the *i*-th node.

The relative load distributed from the *i*-th region, that is, $L_R'(\mathrm{i})$ is defined as

$$L_R'(\mathrm{i}) = \frac{L_R(\mathrm{i})}{I_{R,\max}(\mathrm{i})} \tag{4.14}$$

Where, $I_{\mathrm{R,max}}(\mathrm{i})$ is the load current limitation of the *i*-th region.

Sometimes, the power factor of the loads may be corrected to a rather high level. Therefore, the load through a node or distributed from a region may be expressed in the form of current (A). We have the *current form load descriptions* as

$$l_{\mathrm{N}}(\mathrm{i}) = \mathrm{L}_{\mathrm{N}}(\mathrm{i}) = I_{\mathrm{N}}(\mathrm{i}) \tag{4.15}$$

$$l_{\mathrm{R}}(\mathrm{i}) = \mathrm{L}_{\mathrm{R}}(\mathrm{i}) = I_{\mathrm{R}}(\mathrm{i}) \tag{4.16}$$

4.2.2.2 Node-to-Region Load Transform

Node-to-region load transform describes the program calculating loads distributed from regions according to the loads through the nodes. In other words, it solves the loads distributed from the regions when the loads through the nodes are known.

As for a region with all loads through its terminals known, the load distributed from the region can be obtained by the load through its enter point minus all loads through its terminals.

As an example, given a region Re(N_i,N_j,…,N_k) that has its terminals N_i, N_j,…,N_k, the load distributed from the region Re(N_i,N_j,…,N_k), that is, $l_R(N_i,N_j,\ldots,N_k)$ can be obtained according to the loads through its terminals as shown in Equation (4.17)

$$l_R(N_i, N_j, \ldots, N_k) = l_N(N_i) - \sum_{\lambda \in \Lambda} l_N(\lambda) \tag{4.17}$$

Where, N_i is the enter point of the region Re(N_i,N_j,…,N_k), Λ is the set of the out points of the region, that is, $\Lambda = [N_j,\ldots,N_k]$.

Under normal conditions, the loads through the nodes may be obtained in real time from the telemetering information gathered from the Feeder Terminal Units (FTU) by the master station of a Distribution Automation System (DAS), based on this the master station may calculate the load distributed from the regions by executing node-to-region load transform.

The information about the load distributed from the regions is very important and should be documented. This information indicates the load distribution pattern, which needs consideration for network reconfiguration and service restoration optimization. The trend of the load distributed from each region is of great help when forecasting the load distributed from the corresponding region in future.

4.2.2.3 Region-to-Node Load Transform

Region-to-node load transform describes the program calculating the loads flowing through the nodes according to loads distributed from the regions. In other words, it solves the loads flowing through nodes when loads distributed from the regions are known.

As for a node with all loads distributed from its downstream regions known, the load flowing through the node is the summation of all loads distributed from its downstream regions, that is,

$$l_N(N_i) = \sum_{d \in \mathbf{D}} l_R(d) \tag{4.18}$$

Where, **D** is the set of the i-th node's downstream regions.

If the i-th node is the enter point of the region $\mathrm{Re}(N_i, N_j, \ldots, N_k)$, the load flowing through the i-th node can be also calculated according to Equation (4.19)

$$l_N(N_i) = l_R(N_i, N_j, \ldots, N_k) + \sum_{\lambda \in \Lambda} l_N(\lambda) \tag{4.19}$$

Where, Λ is the set of the out points of the region $\mathrm{Re}(N_i, N_j, \ldots, N_k)$, i.e., $\Lambda = [N_j, \ldots, N_k]$.

As the information about the load distributed has been documented, the network reconfiguration program and service restoration program may evaluate the candidate schemes by executing region-to-node load transform.

The load through a node is influenced by the operation mode, that is, the network configuration, which weakens the regularity of its trend. Thus, forecasting the load through a certain node just based on the history load through the corresponding node is not always satisfied. But the load distributed from a region is not affected by the operation mode. Therefore, forecasting the load distributed from a region is usually satisfied. Thus, based on the forecasted load distributed from the regions and by executing region-to-node load transform may obtain a good forecasting result through the nodes. It is a promising approach to load forecasting for DASs.

4.2.2.4 An Example

The distribution grid shown in Fig. 4.2 is used as an example to illustrate the simplified load flow analysis approach. Assuming that all loads are corrected to a high power factor, for example, 0.95, therefore the loads may be expressed in the form of current (A).

The numbers in Fig. 4.9 indicate the load current (A) flowing through the corresponding nodes, which are gathered by FTUs.

By executing node-to-region load transform, the loads distributed from the regions may be obtained and illustrated in Fig. 4.10 in brackets.

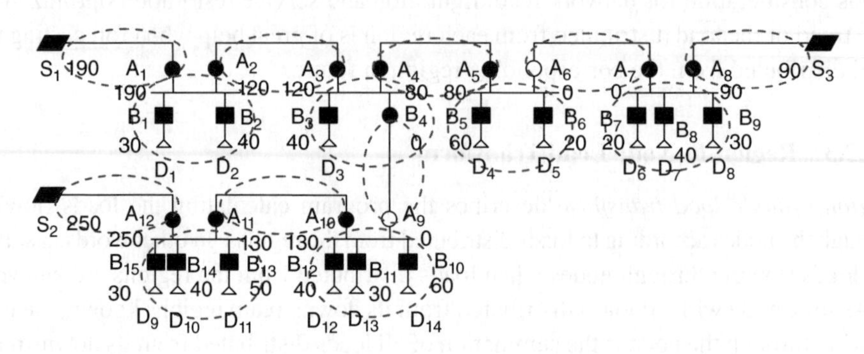

Figure 4.9 The loads flowing through the nodes of the example in Fig. 4.2

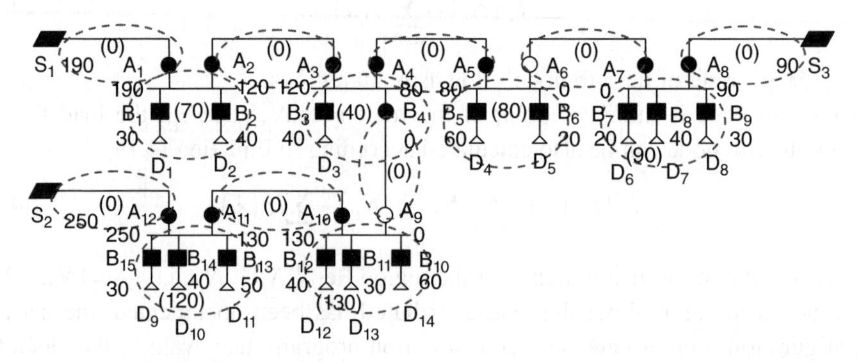

Figure 4.10 The loads distributed from the regions obtained by executing node-to-region load transform

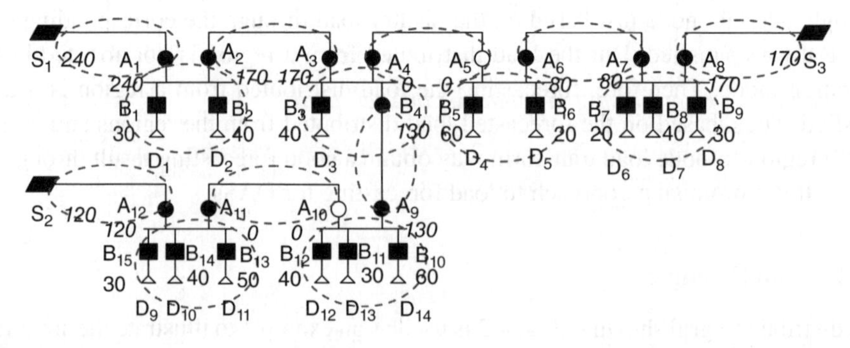

Figure 4.11 The obtained loads through the nodes in a candidate configuration by executing region-to-node load transform

Assuming that the configuration in Fig. 4.11 is a candidate scheme of network reconfiguration, based on the documented loads distributed from the regions and by executing region-to-node load transform, we may obtain the loads flowing through the nodes, which are the italic numbers shown in Fig. 4.11.

4.3 Interphase Short Circuit Fault Location

As described in Section 4.2.1, a monitoring region is the smallest unit of fault location. The criteria described in this section are all monitoring region oriented. They are also all intended for interphase short circuit fault location. The criteria for single-phase grounding fault location in a neutral grounded system are the same as those for interphase short circuit fault location. Problems with single-phase grounding fault location in neutral non effective grounded systems will be discussed separately in Chapter 5.

4.3.1 Fault Location with Sufficient Information

The meaning of *sufficient information for fault location* is that the master station of a DAS receives all of the fault related information from the devices mounted at the corresponding monitoring nodes in time with nothing missing or wrong.

4.3.1.1 Fault Location Criterion for Single Source Open-Loop Distribution Grids – Criterion 4.1

For a single source open-loop distribution grid, once a fault occurs on a feeder section, the fault current may flow through all of the upstream monitoring nodes, but none of the downstream monitoring nodes would receive the fault current.

Thus, Criterion 4.1 is as follows:

For a region in a single source open-loop distribution grid, if its enter point endures the fault current and none of its out points endure the fault current, the fault occurs in the region. Otherwise, the region is healthy. In other words, the fault is in the region with the fault current flowing through just one terminal.

An example is shown in Figure 4.12, in which node-1 is both the source and monitoring node; node-2, node-3, node-4, and node-5 are monitoring nodes; and node-6, node-7, and node-8 are ending nodes. Supposing that a fault occurs in region Re(node-4, node-3,node-5). The marks "↑" beside the nodes indicate that the corresponding nodes endure fault current.

For the region Re(node-1, node-2, node-4), although its enter point (i.e., node-1) endures the fault current, one of its out points (i.e., node-4) also endures the fault current, thus the fault is not in the region.

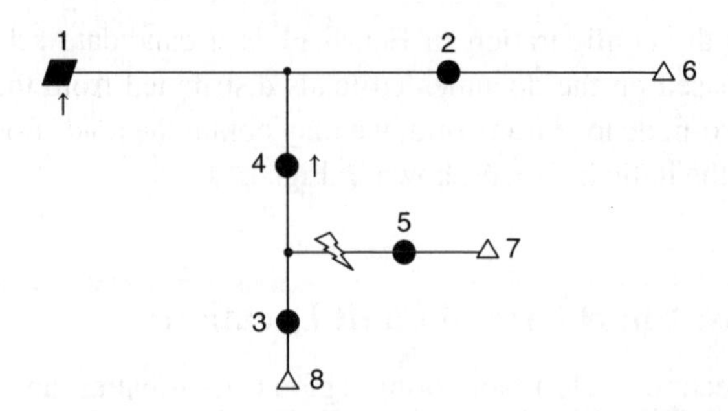

Figure 4.12 An example of fault location for single source open-loop distribution grids

For the region Re(node-2, node-6), since its enter point (i.e., node-2) does not endure the fault current, the fault is not in the region. Similarly, Re(node-5, node-7) and Re(node-3, node-8) are healthy.

For the region Re(node-4, node-3, node-5), its enter point (i.e., node-4) endures the fault current, neither of its out points (i.e., node-3 and node-5) endures the fault current. According to Criterion 4.1, the fault is in the region.

It is worth mentioning that whether or not the current flowing through monitoring nodes is over the threshold set in advance must be detected so as to meet the requirement of Criterion 4.1. The magnitude of the current does not necessarily have to be accurately measured. Neither voltage nor the direction of fault power flow is needed to detect it.

4.3.1.2 Fault Location Criterion for Multi-Main-Source Close-Loop Distribution Grids – Criterion 4.2

For a multi-main-source close-loop distribution grid, once a fault occurs on a feeder section, the fault current may flow through all of the nodes on the trunk route and the nodes on the branches that lead to the fault section. Thus, we cannot locate the fault section according to Criterion 4.1, because sometimes more than one terminal of the fault region may endure the fault current.

In this case, the directions of fault power flows through the monitoring nodes are needed to locate the fault position. The *fault location criterion for multi-main-source close-loop distribution grids*, that is, Criterion 4.2, is as follows:

For a region in a multi-main-source close-loop distribution grid, if at least one terminal endures a fault current with the direction of fault power flow pointing to the interior of the region, and none of the terminals endures the fault current with the direction of fault power flow pointing to the external of the region, the fault occurs in the region. Otherwise, the region is healthy.

An example is shown in Figure 4.13, in which node-1 and node-8 are both source nodes and monitoring nodes. Node-10 is an ending node and the other nodes are monitoring nodes.

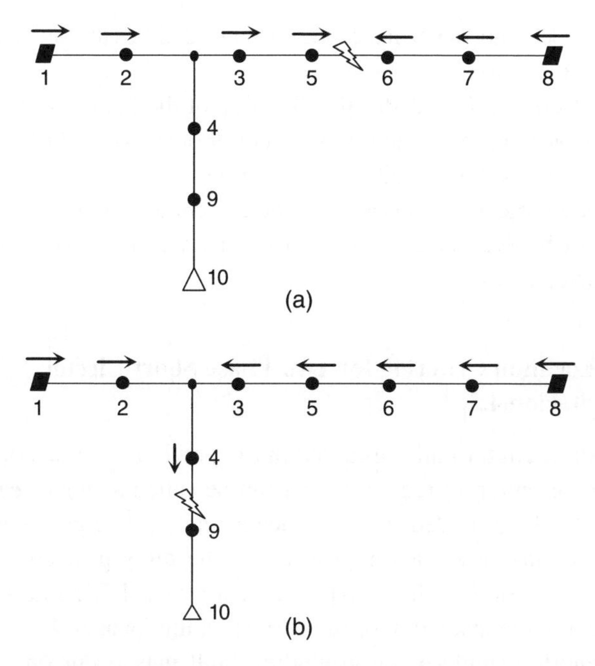

Figure 4.13 An example of fault location for multi-main-source close-loop distribution grids

First, suppose that a fault occurs in region Re(node-5, node-6), which is shown in Figure 4.13(a). The arrows beside the nodes indicate the direction of the fault power flowing through the corresponding nodes.

For the region Re(node-2, node-3, node-4), although the direction of the fault power flowing through node-2 points to the interior of the region, the directions of the fault power flowing through node-3 points to the external of the region, thus the fault is not in the region. Similarly, Re(node-1, node-2), Re(node-3, node-5), Re(node-6, node-7), and Re(node-7, node-8) are healthy.

For the region Re(node-4, node-9), none of its terminals endure the fault current, therefore, the fault is not in the region. Similarly, Re(node-9, node-10) is also healthy.

For the region Re(node-5, node-6), the direction of the fault power flowing through all terminals points to the interior of the region. According to Criterion 4.2, the fault is in the region.

Suppose that a fault occurs in region Re(node-4, node-9), which is shown in Fig. 4.13(b).

For the region Re(node-1, node-2), although the direction of the fault power flowing through node-1 points to the interior of the region, the direction of the fault power flowing through node-2 points to the exterior of the region, thus the fault is not in the region. Similarly, Re(node-2, node-3, node-4), Re(node-3, node-5), Re(node-5, node-6), Re(node-6, node-7), and Re(node-7, node-8) are healthy.

For region Re(node-9, node-10), none of its terminal endures the fault current, therefore, the fault is not in the region.

For the region Re(node-4, node-9), the direction of the fault power flowing through node-4 points to the interior of the region. Node-9 does not endure the fault current. According to Criterion 4.2, the fault is in the region.

To meet the requirement of Criterion 4.2, the current detected flowing through monitoring nodes must be over the threshold set in advance and the direction of the fault power must be detected too.

4.3.1.3 Fault Location Criterion for Two-Phase Short Circuit to Ground – Criterion 4.3

The interphase short circuit fault discussed in the previous sections occurs within the scope of only one monitoring region, which can be called a *single region fault*. For a neutral non-effectively grounded system, when a single-phase grounding fault occurs, the zero-sequence current is too small to cause the relay protection device action. But the phase voltages of the other two phases increase to 1.732 times the rated value, in theory threatening the insulation of the corresponding phases. If the abnormal state lasts too long, another single-phase grounding fault may occur on one of the other phases due to insulation failure. Therefore, a two phase short circuit to ground is initiated with the two grounding positions in different regions, which is called a *multi-region fault*. This kind of fault is quite common in neutral non-effectively grounding systems and the fault positions of these are often difficult to identify. Many DASs don't even have the function of two phase short circuit to ground fault location, which remarkably reduces their contributions to reliability.

For a neutral non-effectively grounded system, once a two phase short circuit to ground fault occurs, the fault phenomena are as follows:

- The zero-sequence voltages on the corresponding bus and on any positions of its connected feeders are over the threshold.
- As for the two grounding phases, the fault current may flow through the upstream monitoring nodes along the two phase-lines, respectively, but none of the downstream monitoring nodes undergo the fault current.
- At least one circuit breaker along the upstream of grounding regions is tripped by a relay protection device.

Figure 4.14(a)–(d) illustrates some typical cases of two phase short circuit to ground faults. The marks "↑" beside the nodes indicate that the corresponding nodes endure over-current or over-voltage. The squares indicate circuit breakers with relay protection devices and the circles indicate load switches. The solid symbols indicate the switches are closed while the hollow symbols indicate the switches are open. The data in brackets indicate the delay time of the relay protection devices.

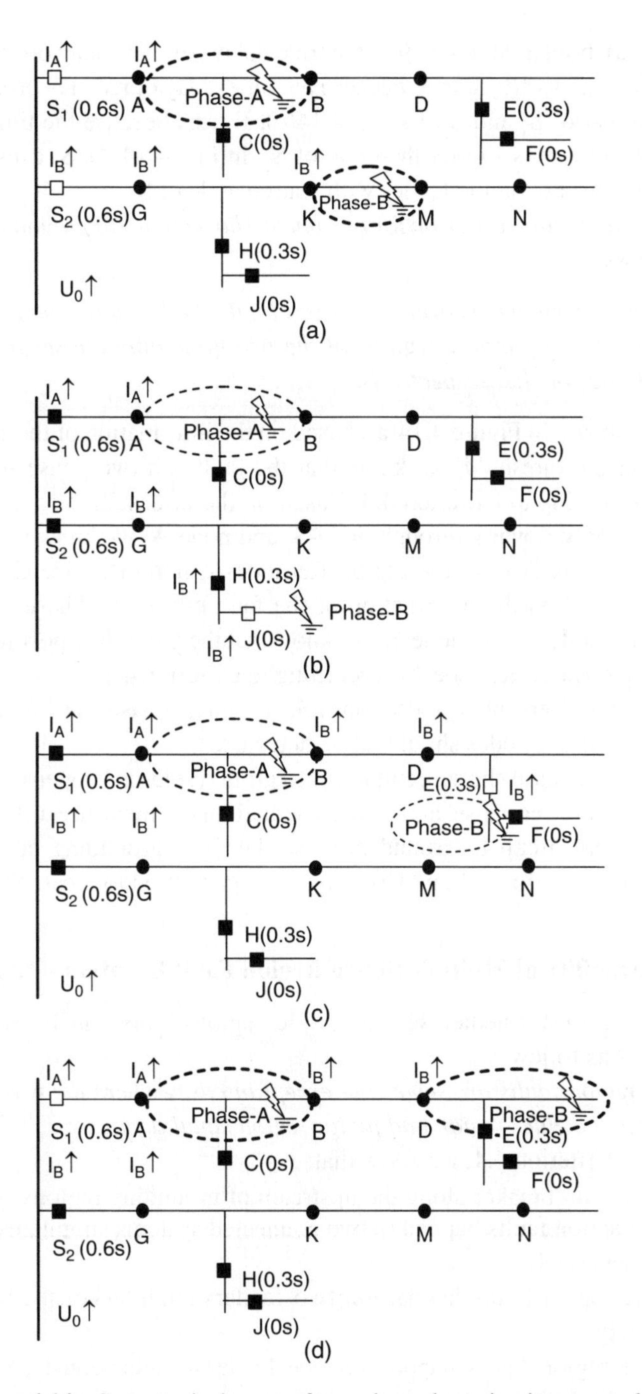

Figure 4.14 Some typical cases of two phase short circuit to ground fault

In Fig. 4.14(a), both S_1 and S_2 trip as the time delays of their relay protection devices are the same. In Fig. 4.14(b), J trips because the time delay of its relay protection device is shorter than that of S_1 and S_2. In Fig. 4.14(c), E trips because the time delay of its relay protection device is shorter than that of S_1. In Fig. 4.14(d), S_1 trips because two grounding positions are on the feeder with source node of S_1.

Thus, the *fault location criterion for two phase short circuit to ground* (i.e., Criterion 4.3) is as follows:

If the zero-sequence voltage is over the threshold, the fault is a two phase short circuit to ground fault. The grounding regions of the two grounding phases may be located according to Criterion 4.1, respectively.

For the case shown in Figure 4.14(a), based on the information of the zero-sequence voltage U_0 over the threshold, we know that the fault is a two phase short circuit to ground fault according to Criterion 4.3. Based on the information from the fault current of Phase-A only flowing through node-S_1 and node-A, we know that the grounding position of Phase-A is in the region Re(node-A, node-B, node-C) according to Criterion 4.1. Based on the information of the fault current of Phase-B flowing just through node-S_2, node-G, and node-K, we know that the grounding position of Phase-B is in the region Re(node-K, node-M) according to Criterion 4.1.

To meet the requirement of Criterion 4.3, the three phase fault currents flowing through the monitoring nodes should all be detected.

The case of three phase short circuit faults to ground seldom happens except in cases of facility failure, pole collapse, and so on. The fault phenomena are different from that of two phase short circuit to ground because the three grounding positions are the same. Thus, both Criterion 4.1 and Criterion 4.3 can be used to locate the fault.

4.3.1.4 Separability of Multiple Single Region Fault Location – Criterion 4.4

The criterion clarifies whether several single region faults can be separated (i.e., Criterion 4.4) is as follows:

Two single region faults are separable except in cases where the two single region faults happen in the elder region and junior region simultaneously.

According to Criterion 4.4, we know that:

- If two single region faults happen in two connected systems simultaneously, we can locate them separately.
- If two single region faults happen on two feeders simultaneously, we can locate them separately.
- If two single region faults happen on one feeder simultaneously, we can locate them separately under the conditions that one of the two fault regions is not the elder region of the other fault region. Otherwise, the two faults cannot be located separately.

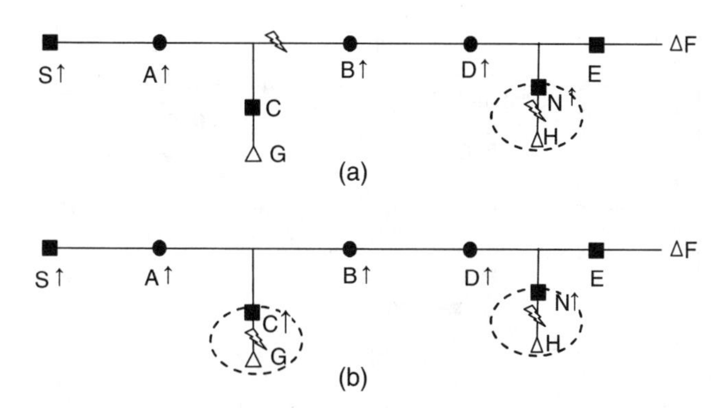

Figure 4.15 Examples to illustrate the separability of multiple single region fault location

Two examples are shown in Fig. 4.15 in which the arrows indicate that the fault current flows through the corresponding nodes.

In Fig. 4.15(a), two faults occurs in Re(A, B, C) and Re(N, H) simultaneously. Re(A, B, C) is the elder region of Re(N, H) (i.e., Re(N, H) is the junior region of Re(A, B, C)). According to Criterion 4.4, the two faults cannot be located separately. In fact, the fault current may flow through node-S, node-A, node-B, node-D, and node-N. According to Criterion 4.1, the fault location result is only Re(N, H).

In Fig. 4.15(b), two faults occur in Re(C, G) and Re(N, H) simultaneously. Re(C,G) and Re(N, H) do not have the elder and junior region relationship. According to Criterion 4.4, the two faults can be located separately. In fact, the fault current flows through node-S, node-A, node-B, node-C, node-D, and node-N. According to Criterion 4.1, the fault location results are Re(C,G) and Re(N, H).

4.3.1.5 Separability of Multiple Multi-Region Faults Location – Criterion 4.5

The criterion to clarify whether several multiple multi-region faults can be separated (i.e., Criterion 4.5) is as follows:

Two multiple multi-region faults are separable if and only if the grounding regions of each phase are separable, which may be determined according to Criterion 4.4.

Some examples are shown in Fig. 4.16 in which the arrows indicate that the fault current flows through the corresponding nodes.

In Fig. 4.16(a), grounding positions of Phase-A are in two regions (i.e., Re(A, B, C) and Re(D, F, P)) simultaneously. The grounding position of Phase-B is in Re(K, M). Re(A, B, C) is the elder region of Re(D, F, P) (i.e., Re(D, F, P) is the junior region of Re(A, B, C)). According to Criterion 4.5, the two grounding positions of Phase-A cannot be located separately. In fact, the Phase-A fault current flows through node-S_1, node-A, node-B, and node-D. According to Criterion 4.1, the grounding region of

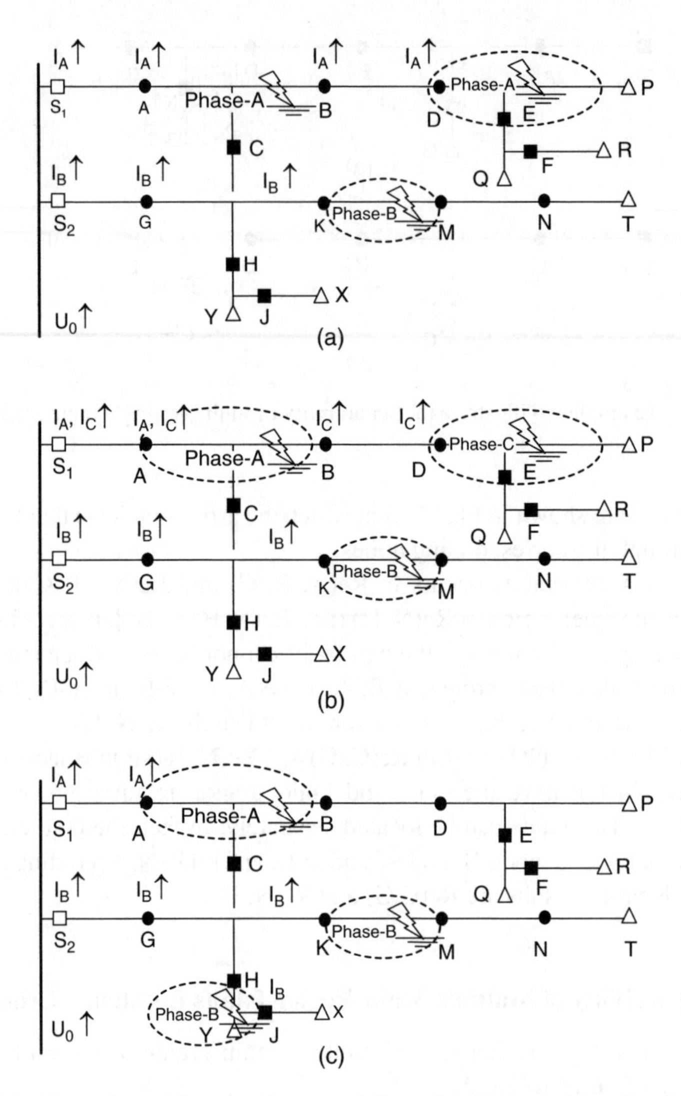

Figure 4.16 Examples to illustrate the separability of multiple multi-region faults location

Phase-A can only be located in Re(D, F, P). The grounding region of Phase-B can be located in Re(K, M).

In Fig. 4.16(b), the grounding position of Phase-A is in Re(A, B, C). The grounding position of Phase-B is in Re(K,M) and the grounding position of Phase-C is in Re(D, E, P) simultaneously. The grounding regions of Phase-A, Phase-B, and Phase-C do not have the elder and junior region relationship. According to Criterion 4.5, the three grounding regions can be located separately. In fact, the fault current of Phase-A flows through node-S_1 and node-A, the fault current of Phase-B flows through node-S_2 and node-G and node-K, and the fault current of Phase-C flows through node-S_1 and node-A, node-B, and node-D. According to Criterion 4.3, the grounding regions of Phase-A, Phase-B and Phase-C are Re(A, B, C), Re(D, E, P) and Re(K, M), respectively.

In Fig. 4.16(c), grounding position of Phase-A is in Re(A, B, C). The grounding positions of Phase-B are in Re(K,M) and Re(H,Y, J) simultaneously. The grounding regions of Phase-B do not have the relationship of elder region and junior region. According to Criterion 4.5, the three grounding regions can be located separately. In fact, the fault current of Phase-A flows through node-S_1 and node-A, the fault current of Phase-B flows through node-S_2 and node-G, node-H and node-K. According to Criterion 4.3, the grounding region of Phase-A is Re(A, B, C). The grounding regions of Phase-B are Re(K, M) and Re(H,Y,J), respectively.

4.3.2 Fault Location with Insufficient Information

Since the FTUs and communication systems are mounted outdoors in severe environmental conditions, such as high temperature fluctuations, high humidity (even torrential rain), and so on, as a result some of the fault related information may sometimes be missed or wrong.

Despite this, the fault may be correctly located since the fault information is often redundant.

4.3.2.1 Redundant Fault Information

The following shows typical cases in which fault information is redundant:

- For single region faults, when an interphase short circuit fault occurs, the fault information of at least two phases may be produced but just one of them is enough.
- For overhead feeders, the reclosing process is used to quickly restore the service in the cases of temporary faults. But, reclosing fails in the cases of permanent faults and all fault information is produced again.
- If fault location is incorrect due to insufficient fault information, the fault will happen again in the restoration process and so all fault information is produced again.
- For FTUs with relay protection performance, once a fault occurs downstream of a switch, the corresponding FTU may trip the switch and send the fault information to the master station,as well. Thus, the information about the tripping of the switch can also be regarded as redundant.

4.3.2.2 The Bayesian Analysis-Based Fault Location Approach

Bayesian analysis is an effective methodology to deal with uncertainty. Thus it is often used to solve problems of fault location with insufficient information.

When we receive a fault information, assuming that the probability of it being correct is α, then the probability of it being mistake is $1-\alpha$.

When we do not receive a fault information, assuming that the probability of it being missing is 1–β, then the probability of it being correct is β.

Generally, the probability of having missing or wrong fault information is much smaller than in correct cases, which is called *small disturbance assumption*, that is,

$$\alpha >> 1-\alpha \tag{4.20}$$

$$\beta >> 1-\beta \tag{4.21}$$

The fault location results are sound when a small disturbance assumption is satisfied.

The Bayesian analysis-based fault location approach compares matching degrees of the gathered fault information to the fault assumptions of all possible regions, based on which, the fault assumption with the largest matching degree (as well as the ones with far greater matching degrees) are pointed out. A fault assumption is a hypothesis of the fault position, such as the fault is in region-6, the grounding position of phase-A is in region-12, while the grounding position of phase-C is in region-36, and so on.

When a fault information of the *i*-th node is received, suppose that the probability of it matching the fault assumption is $p_{C,i}$, we have:

$$p_{C,i} = \alpha_i \tag{4.22a}$$

in cases where the fault information of the *i*-th node is received and matches the fault assumption and

$$p_{C,i} = 1-\alpha_i \tag{4.22b}$$

in the cases where the fault information of the *i*-th node is received but does not match the fault assumption.

When we do not receive any fault information of the *i*-th node, the probability of the phenomenon matching the fault assumption is

$$p_{C,i} = \beta_i \tag{4.23a}$$

in cases it matches the fault assumption and

$$p_{C,i} = 1-\beta_i \tag{4.23b}$$

in cases it does not match the fault assumption.

The probability of the gathered fault information matrix $\boldsymbol{G}_0$ matching the *j*-th fault assumption is

$$p(H_j) = p(\mathrm{G}_0|H_j) = \prod_{i\in\Omega} p_{C,i} \tag{4.24a}$$

Where, Ω is the set of the nodes on the corresponding feeders with fault.

When there are many sets of independent fault information, the probabilities of them to match the *j*-th fault assumption are $p_1(H_j)$, $p_2(H_j)$, and so on, respectively.

Combining these results, we have the probability of these fault information to match the j-th fault assumption is,

$$p(H_j) = p_1(H_j) \times p_2(H_j) \times \ldots \times p_k(H_j) \tag{4.24b}$$

Where, K is the number of fault information sets.

The probability of the j-th fault assumption is

$$P(H_j) = \frac{p(H_j)}{\sum\limits_{i \in \Psi} p(H_i)} \tag{4.25}$$

Where, Ψ is the set of all of the fault assumptions.

We may set thresholds P_{set1} and P_{set2}: if the largest probability is larger than P_{set1}, we consider the corresponding fault assumption is the only fault location result. Otherwise, we consider all the fault assumptions with probability higher than P_{set2} as possible fault location results.

4.3.2.3 Examples

In this section, several examples are given to illustrate Bayesian analysis-based fault location with insufficient information.

For the distribution grid shown in Fig. 4.17(a), S_I and S_{II} are source nodes, which are circuit breakers. A, B, C, D and E are monitoring nodes, which are load switches. R_1, R_2, R_3, R_4, R_5, and R_6 are regions. The solid symbols indicate the corresponding switches are closed. The hollow symbols indicate the corresponding switches are open. Assume that α is 0.9 and β is 0.8, P_{set1} and P_{set2} are set as 0.8 and 0.1, respectively.

First, we investigate the case where a fault with a short circuit between Phase-A and Phase-B occurs in region R_5 causing S_I to trip, as shown in Fig. 4.17(b).

Assume that the gathered fault information $\boldsymbol{G}_0$ is $g_{0,A,SI}$=1, $g_{0,B,SI}$=1; $g_{0,A,A}$=0, $g_{0,B,A}$=0; $g_{0,A,B}$=1, $g_{0,B,B}$=1; $g_{0,A,C}$=1, $g_{0,B,C}$=1; and $g_{0,A,D}$=0, $g_{0,B,D}$=0. As for $g_{0,x,y}$, x indicates the phases, such as Phase-A, Phase-B, and Phase-C; y indicates the node number; $g_{0,x,y}$=1 indicates that the master station receives the information of the fault current of Phase-x flowing through the Node-y; and $g_{0,x,y}$=0 indicates that the master station does not receive the fault information from Node-y.

It can be seen from the gathered fault information that contradictions exist between node-A and node-B. Since node-A is the father node of node-B, if the fault current flows through node-B, it should also flow through node-A. But the master station does not receive the fault information from node-A thus making the fault information insufficient. The reasons for this phenomenon may be interrupted communication, FTU destroyed, current transformer damage, and so on.

Based on this fault information, as for a poor DAS, which is not able to deal with insufficiency information, cannot locate the fault position or mistakenly locate the fault in region R_2 according to Criterion 4.1.

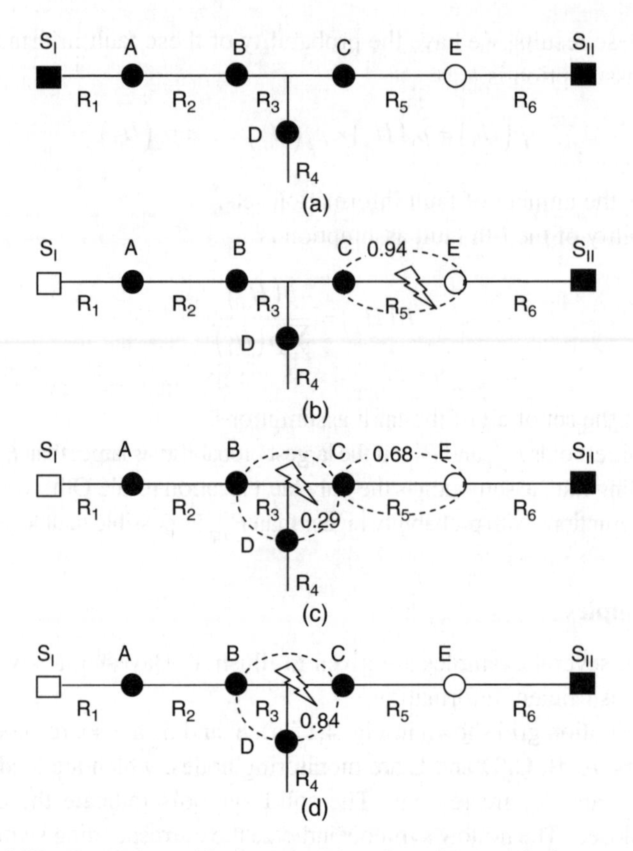

Figure 4.17 Examples to illustrate fault location with insufficient information

As for the fault assumption of the fault being in R_5, we have

$$p_{C,A,SI} = \alpha = 0.9,\ p_{C,B,SI} = \alpha = 0.9$$

$$p_{C,A,A} = 1-\beta = 0.2,\ p_{C,B,A} = 1-\beta = 0.2$$

$$p_{C,A,B} = \alpha = 0.9,\ p_{C,B,B} = \alpha = 0.9$$

$$p_{C,A,C} = \alpha = 0.9,\ p_{C,B,C} = \alpha = 0.9$$

$$p_{C,A,D} = \beta = 0.8,\ p_{C,B,D} = \beta = 0.8$$

According to Equation (4.24), we have

$$p(R_5) = 0.9^2 \times 0.2^2 \times 0.9^4 \times 0.8^2 = 1.4 \times 10^{-2}$$

Similarly, we have

$$p(R_0) = 0.1^2 \times 0.8^2 \times 0.1^4 \times 0.8^2 = 4.1 \times 10^{-7}$$

$$p(R_1) = 0.9^2 \times 0.8^2 \times 0.1^4 \times 0.8^2 = 3.3 \times 10^{-5}$$

$$p(R_2) = 0.9^2 \times 0.2^2 \times 0.1^4 \times 0.8^2 = 2.1 \times 10^{-6}$$

$$p(R_3) = 0.9^2 \times 0.2^2 \times 0.9^2 \times 0.1^2 \times 0.8^2 = 1.6 \times 10^{-4}$$

$$p(R_4) = 0.9^2 \times 0.2^2 \times 0.9^2 \times 0.2^2 \times 0.8^2 = 6.7 \times 10^{-4}$$

where, R_0 indicates no fault occurs.

According to Equation (4.25), we have

$$P(R_0) \approx 0,\ P(R_1) \approx 0,\ P(R_2) \approx 0,\ P(R_3) \approx 0.01,\ P(R_4) \approx 0.05,\ P(R_5) \approx 0.94$$

Because the probability of the assumption of the fault being in region R_5 is larger than P_{set1} (i.e., 0.8), the fault location result is only region R_5, which is shown in Fig. 4.17(b).

It can be seen from these fault location results that, although the fault information is insufficient, the fault region can be uniquely and correctly located in region R_5.

Second, we investigate the case where a permanent fault with short circuit between Phase-A and Phase-B occurs in region R_3 causing S_I to trip.

Assume that the gathered fault information $\boldsymbol{G}_0$ is $g_{0,A,SI}$=1, $g_{0,B,SI}$=1; $g_{0,A,A}$=1, $g_{0,B,A}$=0; $g_{0,A,B}$=1, $g_{0,B,B}$=1; $g_{0,A,C}$=1, $g_{0,B,C}$=0; and $g_{0,A,D}$=0, $g_{0,B,D}$=0. It can be seen from the gathered fault information that contradictions exist for node-A and node-C so there is little valuable information about the fault.

For some DAS with insufficient information about the fault location, the contradictions are ignored. Once the fault information is received for only one phase, the corresponding node is marked with the fault current flowing through it. Thus, in this example, the fault current is marked flowing through node-S_I, node-A, node-B, and node-C, based on which the fault position is mistakenly located in region R_5 according to Criterion 4.1.

As for the fault assumption in R_5, we have

$$p_{C,A,SI} = \alpha = 0.9,\ p_{C,B,SI} = \alpha = 0.9$$

$$p_{C,A,A} = \alpha = 0.9,\ p_{C,B,A} = 1-\beta = 0.2$$

$$p_{C,A,B} = \alpha = 0.9,\ p_{C,B,B} = \alpha = 0.9$$

$$p_{C,A,C} = \alpha = 0.9,\ p_{C,B,C} = 1-\beta = 0.2$$

$$p_{C,A,D} = 1-\beta = 0.8,\ p_{C,B,D} = 1-\beta = 0.8$$

According to Equation (4.24), we have

$$p(R_5) = 0.9^3 \times 0.2 \times 0.9^3 \times 0.2 \times 0.8^2 = 1.4 \times 10^{-2}$$

Similarly, we have

$$p(R_0) = p_0(R_0) = 0.1^3 \times 0.8 \times 0.1^3 \times 0.8^3 = 4.1 \times 10^{-7}$$

$$p(R_1) = p_0(R_1) = 0.9^2 \times 0.1 \times 0.8 \times 0.1^3 \times 0.8^3 = 3.3 \times 10^{-5}$$

$$p(R_2) = p_0(R_2) = 0.9^2\times0.9\times0.2\times0.1^3\times0.8^3 = 7.5\times10^{-5}$$

$$p(R_3) = p_0(R_3) = 0.9^2\times0.9\times0.2\times0.9^2\times0.1\times0.8^3 = 6.0\times10^{-3}$$

$$p(R_4) = p_0(R_4) = 0.9^2\times0.9\times0.2^3\times0.9^2\times0.1\times0.8 = 3.8\times10^{-4}$$

According to Equation (4.25), we have

$$P(R_0)\approx0,\ P(R_1)\approx1.6\times10^{-3},\ P(R_2)\approx3.7\times10^{-3},\ P(R_3)\approx0.29,\ P(R_4)\approx0.02,\ P(R_5)\approx0.68$$

Because none of the fault assumptions show a probability larger than P_{set1} (i.e., 0.8), the fault location results are not unique. Comparing with P_{set2} (i.e., 0.1), both R_5 and R_3 are the possible fault regions with the possibility of R_5 being much higher than R_3, which is shown in Fig. 4.17(c).

It can be seen from these fault location results that, although the fault region is not uniquely and correctly located in region R_3, the possibility of R_3 increases for consideration while the possibility of R_5 decreases remarkably to a relatively low level.

The confusion of fault location is because there is too much wrong fault information.

For the overhead feeder, reclosing is permitted. But reclosing fails due to fact the fault is permanent, causing source node S_I to trip again. The fault information gathered during the reclosing process may improve the quality of fault location. Assume that the gathered fault information $\boldsymbol{G}_R$ is $g_{R,A,SI}=1$, $g_{R,B,SI}=1$; $g_{R,A,A}=1$, $g_{R,B,A}=0$; $g_{R,A,B}=1$, $g_{R,B,B}=1$; $g_{R,A,C}=0$, $g_{R,B,C}=0$; and $g_{R,A,D}=0$, $g_{R,B,D}=0$.

It can be seen from $\boldsymbol{G}_R$ that the contradiction only exists for node-A. The current transformer mounted on phase-B of node-A may be broken.

According to Equation (4.24), we have

$$p_R(R_0) = 0.1^3\times0.8\times0.1^2\times0.8^4 = 3.2\times10^{-6}$$

$$p_R(R_1) = 0.9^2\times0.1\times0.8\times0.1^2\times0.8^4 = 2.1\times10^{-4}$$

$$p_R(R_2) = 0.9^2\times0.9\times0.2\times0.1^2\times0.8^4 = 4.8\times10^{-4}$$

$$p_R(R_3) = 0.9^2\times0.9\times0.2\times0.9^2\times0.8^4 = 4.8\times10^{-2}$$

$$p_R(R_4) = 0.9^2\times0.9\times0.2\times0.9^2\times0.2^2\times0.8^2 = 3.0\times10^{-3}$$

$$p_R(R_5) = 0.9^2\times0.9\times0.2\times0.9^2\times0.2\times0.2\times0.8^2 = 3.0\times10^{-3}$$

Combining the fault information before and after reclosing together, we have

$$p(R_0) = p_0(R_0)p_R(R_0) = 4.1\times10^{-7}\times3.2\times10^{-6}\approx1.3\times10^{-12}$$

$$p(R_1) = p_0(R_1)p_R(R_1) = 3.3\times10^{-5}\times2.1\times10^{-4}\approx6.9\times10^{-9}$$

$$p(R_2) = p_0(R_2)p_R(R_2) = 7.5\times10^{-5}\times4.8\times10^{-4}\approx3.6\times10^{-8}$$

$$p(R_3) = p_0(R_3)p_R(R_3) = 6.0\times10^{-3}\times4.8\times10^{-2}\approx2.9\times10^{-4}$$

$$p(R_4) = p_0(R_4)p_R(R_4) = 3.8\times10^{-4}\times3.0\times10^{-3}\approx1.14\times10^{-6}$$

$$p(R_5) = p_0(R_5)p_R(R_5) = 1.4\times10^{-2}\times3.0\times10^{-3}\approx5.2\times10^{-5}$$

According to Equation (4.25), we have

$$P(R_0)\approx 0,\ P(R_1)\approx 0,\ P(R_2)\approx 0,\ P(R_3)\approx 0.84,\ P(D_4)\approx 0.01,\ P(D_5)\approx 0.15$$

Because the probability of the assumption of the fault being in region R_3 is larger than P_{set1} (i.e., 0.8), the fault location result shows only region R_3, which is shown in Fig. 4.17(d).

4.3.3 Fault Location for Distribution Grids with DGs

DGs may have influence on fault location since they may not only change the amplitude of the fault current and the direction of the fault power flow, but also they may change the distribution pattern of the fault current, sometimes making a distribution grid like a radial grid, and others like a close-loop grid.

The extent of influence of the different types of DG on fault location varies a lot. The machine type DGs with large power ratings may have remarkable influence while the inverter type DGs seldom cause problems.

The access positions of DGs also affect the extent of the influence on fault location. The DGs on a feeder have some influence while the DGs directly attached to the buses usually have no influence on fault location, which will not discussed in this section. The difference between the main side short circuit capacities also affects the influence of DGs on fault location.

The fault location criteria and methodologies described in Sections 4.3.1 and 4.3.2 are called *traditional fault location approaches*, which apply to the distribution grids without DGs. But they are still suitable for solving the problems of fault location for most distribution grids with DGs.

4.3.3.1 The Influence of DGs on Traditional Fault Location Approaches

Here, we discuss the situations of DGs attached to feeders.

For multi-main-source close-loop distribution grids with DGs, the DGs have no influence on fault location according to Criterion 4.2. Thus, we will not discuss it in the following.

For single source open-loop distribution grids with DGs, when a fault occurs, the fault current from the main source not only flows through upstream nodes, but the fault current from DGs may also flow through downstream nodes. If the fault current from the DGs is really strong, the fault information reporting thresholds of FTUs are exceeded. The master station of the DAS may receive the fault information not only from the enter point of the fault region but also from some out points. Thus, the fault may not be correctly located according to Criterion 4.1 or Criterion 4.3. Criterion 4.3 is based on Criterion 4.1; thus if Criterion 4.1 is effective then so is Criterion 4.3.

For a single source open-loop distribution grid with DGs, if the fault current from the main source is always much larger than the fault current from the DGs, we may set an appropriate threshold of I_{set} to report just the fault current from the main source. As a result, the fault information is the same as the distribution grid without DGs. Therefore, the fault region may be correctly located according to Criterion 4.1. The key is to guarantee the constant threshold is always effective, which requires us to find out the worst condition. Once the threshold is effective in the worst conditions, it is always effective.

4.3.3.2 The Worst Condition

The worst condition of a radial feeder is as follows:

- A short circuit fault between only two phases occurs on the furthest end of the feeder, in which case, the fault current from the main is only 0.866 times that of the three phase short circuit, but the fault current from inverter-type DGs does not change.
- The decrease in fault current from the main source due to DGs is considered as the sum of the fault current from all DGs on the feeder.
- The fault current from each DG reaches its maximum with sufficient primary power.

The worst condition of a looped feeder pair is as follows:

- The operation mode is that the loop switch is closed; that is, a feeder serves not only its own load but also the load of the feeder connected with it through the loop switch.
- A short circuit fault between only two phases occurs on the furthest end of the feeder pair with the loop switch closed.
- The decrease in fault current from the main source due to DGs is considered as the sum of the fault current from all of the DGs on the feeder and the feeder connected with it through the loop switch.
- The fault current from each DG reaches its maximum with sufficient primary power.

It can be seen that the worst conditions do not exist in practice but this is more stringent than any situation in reality.

In the worst condition, the minimum fault current from the main source, that is, $I''_{k.min}$, is

$$I''_{k.min} = 0.5\sqrt{3}I_{k.min} - I_{DG} \tag{4.26}$$

Where, I_{DG} is the total fault current from the DGs on one feeder or the two feeders connected with the closed loop switch and $I_{k.min}$ is the three phase short circuit current on the furthest end of the feeder or feeder pair with the loop switch closed, which may be calculated by short circuit current calculation approaches.

4.3.3.3 The Effective Scope of Criterion 4.1

Assuming that the rated current of a feeder is I_C. No matter what the conditions are, the loading current of a feeder should not exceed I_C.

4.3.3.3.1 Distribution Grids with Only Inverter-Type DGs

To investigate the case of only inverter-type DGs, we assume that the ratio of total capacity of inverter-type DGs to the rated current of the feeder is γ. According to [43], the maximum fault current from an inverter-type DG is no more than 1.5 times of its rated current. Thus, in the worst condition, all inverter-type DGs on a radial feeder provide the maximum fault current of $1.5\gamma I_C$. For a looped feeder pair in the worst condition, the fault current may come from the DGs on the two feeders connected with the closed loop switch, which is double of the case of a radial feeder. Therefore, the maximum fault current from inverter-type DGs, that is, I_{DG} is

$$I_{DG} = \begin{cases} 1.5\gamma I_C & (for \quad radial \quad feeder) \\ 3\gamma I_C & (for \quad loop \quad feeders) \end{cases} \tag{4.27}$$

Taking a parameter for reliability K_k into consideration, we may set the threshold I_{set} of the fault information reporting as $K_k I_{DG}$. Typically, K_k is a value from 2 to 3. When $K_k = 2$, we have

$$I_{SET} = K_k I_{DG} = \begin{cases} 2\times(1.5\gamma I_C) = 3\gamma I_C & (for \quad radial \quad feeder) \\ 2\times(2\times 1.5\gamma I_C) = 6\gamma I_C & (for \quad loop \quad feeders) \end{cases} \tag{4.28}$$

If Equation (4.28) is satisfied, the fault currents from the DGs are not reported. Thus, Criterion 4.1 is effective when

$$I''_{k.min} > I_{SET} \tag{4.29}$$

If Equation (4.29) is satisfied, the fault current from the main source may be reported making the fault information the same as for a distribution grid without DGs.

Putting Equations (4.26), (4.27), and (4.28) into (4.29), we have

$$\begin{cases} I_{\text{k.min}} > 3\sqrt{3}\gamma \ \ I_{\text{C}} \ \ (for \ \ radial \ \ feeder) \\ I_{\text{k.min}} > 6\sqrt{3}\gamma \ \ I_{\text{C}} \ \ (for \ \ loop \ \ feeders) \end{cases} \tag{4.30}$$

Since $I_{\text{k.min}}$ is determined by the main side short circuit capacities, the specification of the feeder, and the service length of the feeder, according to Equation (4.30) and the three phase short current calculation, we may obtain the maximum load reach of overhead feeders and cable feeders with various specifications to guarantee Criterion 4.1 effective when the capacity of inverter-type DGs is given. Note that, as for a looped feeder pair, the load reach is the sum of the load reaches of the two feeders.

Some load reaches that guarantee Criterion 4.1 effective are shown in Tables 4.2–Table 4.5, respectively.

Similarly, when the load reach is given, we may obtain $I_{\text{k.min}}$ according to the three phase short current calculation, based on which the capacity limitations of inverter-type DGs to guarantee Criterion 4.1 effective, that is, $\gamma_{\text{max}} I_{\text{C}}$ can be calculated according to Equation (4.30) with the constraint condition of the rated current of the feeder.

Some capacity limitations of inverter-type DGs to guarantee Criterion 4.1 effective are shown in Tables 4.6 and Table 4.7, respectively.

It can be seen from Tables 4.2–4.7 that the adaptability of cables is much higher than that of overhead feeders in order to guarantee Criterion 4.1 effective. The capacity limitations of inverter-type DGs decrease with the increasing load reach. It can be found that in most cases Criterion 4.1 may be effective for distribution grids with inverter-type DGs, no matter whether they are overhead feeders or cables.

Table 4.2 Load reach limitations to guarantee Criterion 4.1 effective for a radial feeder (km) (Inverter type DGs with $\gamma = 25\%$)

Type	specification	Rated current (A)	DG capacity (MW)	Main side short circuit capacities (MVA)		
				100	200	500
Overhead feeder	LGJ-95	295	1.28	31.0	31.9	32.4
	LGJ-120	335	1.45	29.3	30.3	30.9
	LGJ-150	391	1.69	26.0	27.1	27.8
	LGJ-185	453	1.96	23.3	24.6	25.3
	LGJ-240	536	2.32	19.8	21.1	21.9
	LGJ-300	615	2.66	17.2	18.5	19.3
Cable feeder	YJV-95	298	1.29	69.4	70.4	71.0
	YJV-120	341	1.48	72.6	74.1	74.9
	YJV-150	390	1.69	73.3	75.3	76.5
	YJV-185	444	1.92	73.0	75.7	77.2
	YJV-240	523	2.26	70.6	74.1	76.2
	YJV-300	610	2.64	64.6	68.8	71.2

Table 4.3 Load reach limitations to guarantee Criterion 4.1 effective for a radial feeder (km) (Inverter type DGs with $\gamma = 50\%$)

Type	specification	Rated current (A)	DG capacity (MW)	Main side short circuit capacities (MVA)		
				100	200	500
Overhead feeder	LGJ-95	295	2.55	14.6	15.5	16.0
	LGJ-120	335	2.90	13.6	14.6	15.2
	LGJ-150	391	3.39	11.9	13.0	13.7
	LGJ-185	453	3.92	10.4	11.7	12.4
	LGJ-240	536	4.64	8.6	9.9	10.7
	LGJ-300	615	5.33	7.3	8.6	9.4
Cable feeder	YJV-95	298	2.58	33.6	34.7	35.3
	YJV-120	341	2.95	34.7	36.3	37.2
	YJV-150	390	3.38	34.5	36.7	37.9
	YJV-185	444	3.85	33.7	36.5	38.1
	YJV-240	523	4.53	31.6	35.3	37.4
	YJV-300	610	5.28	27.9	32.3	34.8

Table 4.4 Load reach limitations to guarantee Criterion 4.1 effective for looped feeders (km) (Inverter type DGs with $\gamma = 25\%$)

Type	specification	Rated current (A)	DG capacity (MW)	Main side short circuit capacities (MVA)		
				100	200	500
Overhead feeder	LGJ-95	295	1.28	14.6	15.5	16.0
	LGJ-120	335	1.45	13.6	14.6	15.2
	LGJ-150	391	1.69	11.9	13.0	13.7
	LGJ-185	453	1.96	10.4	11.7	12.4
	LGJ-240	536	2.32	8.6	9.9	10.7
	LGJ-300	615	2.66	7.3	8.6	9.4
Cable feeder	YJV-95	298	1.29	33.6	34.7	35.3
	YJV-120	341	1.48	34.7	36.3	37.2
	YJV-150	390	1.69	34.5	36.7	37.9
	YJV-185	444	1.92	33.7	36.5	38.1
	YJV-240	523	2.26	31.6	35.3	37.4
	YJV-300	610	2.64	27.9	32.3	34.8

Table 4.5 Load reach limitations to guarantee Criterion 4.1 effective for looped feeders (km) (Inverter type DGs with $\gamma = 50\%$)

Type	specification	Rated current (A)	DG capacity (MW)	Main side short circuit capacities (MVA)		
				100	200	500
Overhead feeder	LGJ-95	295	2.55	6.4	7.3	7.8
	LGJ-120	335	2.90	5.8	6.8	7.4
	LGJ-150	391	3.39	4.8	5.9	6.6
	LGJ-185	453	3.92	3.9	5.2	6.0
	LGJ-240	536	4.64	3.0	4.3	5.1
	LGJ-300	615	5.33	2.3	3.6	4.4
Cable feeder	YJV-95	298	2.58	15.5	16.8	17.5
	YJV-120	341	2.95	15.5	17.4	18.3
	YJV-150	390	3.38	14.8	17.3	18.5
	YJV-185	444	3.85	13.7	16.9	18.5
	YJV-240	523	4.53	11.7	15.8	18.0
	YJV-300	610	5.28	9.2	14.0	16.6

Table 4.6 Capacity limitations of inverter-type DGs to guarantee Criterion 4.1 effective for cable feeders

Load reach (km)	Main side short circuit capacities (MVA)	Capacity limitations (MW)		
		YJV-70	YJV-150	YJV-240
2	100	15.06/4.24	16.22/6.75	16.45/9.06
	200	22.63/4.24	27.29/6.75	28.43/9.06
	500	29.81/4.24	44.02/6.75	49.34/9.06
4	100	11.32/4.24	13.65/6.75	14.22/9.06
	200	14.26/4.24	20.08/6.75	22.04/9.06
	500	16.15/4.24	26.78/6.75	32.06/9.06
6	100	8.81/4.24	11.62/6.75	12.44/9.06
	200	10.22/4.24	15.64/6.75	17.85/9.06
	500	11.02/4.24	19.09/6.75	23.60/9.06
8	100	7.13/4.24	10.04/6.75	11.02/9.06
	200	7.93/4.24	12.73/6.75	14.94/9.06
	500	8.36/4.24	14.80/6.75	18.64/9.06
10	100	5.96/4.24	8.80/6.75	9.87/9.06
	200	6.46/4.24	10.71/6.75	12.82/9.06
	500	6.73/4.24	12.08/6.75	15.40/9.06

Note: the h1 in "h1/h2" indicates the theoretical value of the capacity to guarantee Criterion 4.1 effective, the h2 indicates the allowed capacity restricted by the rated current of the feeder.

Table 4.7 Capacity limitations of inverter-type DGs to guarantee Criterion 4.1 effective for overhead feeders

Load reach (km)	Main side short circuit capacities (MVA)	Capacity limitations (MW)		
		LGJ-70	LGJ-150	LGJ-240
	100	8.00/4.24	9.01/6.77	9.25/9.28
3	200	9.63/4.24	11.58/6.77	12.09/9.28
	500	10.84/4.24	13.88/6.77	14.79/9.28
	100	4.82/4.24	5.79	6.05
6	200	5.31/4.24	6.72	7.13
	500	5.64/4.24	7.42/6.77	7.99
	100	3.43	4.25	4.49
9	200	3.66	4.73	5.06
	500	3.81	5.06	5.47
	100	2.66	3.36	3.57
12	200	2.79	3.65	3.91
	500	2.88	3.84	4.16
	100	2.17	2.78	2.96
15	200	2.26	2.97	3.19
	500	2.31	3.09	3.35
	100	1.83	2.36	2.53
18	200	1.89	2.50	2.70
	500	1.93	2.59	2.81
	100	1.58	2.06	2.21
21	200	1.63	2.16	2.33
	500	1.66	2.23	2.42

Note: the h1 in "h1/h2" indicates the theoretical value of the capacity to guarantee Criterion 4.1 effective, the h2 indicates the allowed capacity restricted by the rated current of the feeder.

4.3.3.3.2 Distribution Grids with Only Machine-Type DGs

According to [43], the maximum fault current from a machine-type DG is no more than 10 times of its rated current. Thus, in the worst condition, all the machine-type DGs on a radial feeder provide the maximum fault current of $10\gamma I_C$. As for a looped feeder pair, in the worst condition, the fault current may come from the DGs on the two feeders connected with the loop switch closed, which is double the case of a radial feeder. Therefore, the maximum fault current from machine-type DGs (i.e., I_{DG}) is

$$\begin{cases} I_{DG} = 10\gamma I_C & (for\quad radial\quad feeder) \\ I_{DG} = 20\gamma I_C & (for\quad loop\quad feeders) \end{cases} \tag{4.31}$$

Taking a parameter for reliability K_k into consideration, we may set the threshold I_{set} of the fault information reporting as $K_k I_{DG}$. Typically, K_k is a value from 2 to 3. When $K_k = 2$, we have

$$I_{SET} = K_k I_{DG} = \begin{cases} 2\times(10\gamma I_C) = 20\gamma I_C & (for \quad radial \quad feeder) \\ 2\times(2\times 10\gamma I_C) = 40\gamma I_C & (for \quad loop \quad feeders) \end{cases} \tag{4.32}$$

If Equation (4.32) is satisfied, the fault currents from the DGs are not reported. Similar to the case of just inverter-type DGs, we have

$$\begin{cases} I_{k.min} > 20\sqrt{3}\gamma I_C & (for \quad radial \quad feeder) \\ I_{k.min} > 40\sqrt{3}\gamma I_C & (for \quad loop \quad feeders) \end{cases} \tag{4.33}$$

We may obtain the maximum load reach of overhead feeders and cable feeders with various specifications to guarantee Criterion 4.1 effective when the capacity of machine-type DGs is given, which are shown in Tables 4.8–4.11, respectively. Of course, as for a looped feeder pair, the load reach is the sum of load reaches of the two feeders.

When the load reach is given, the capacity limitations of machine-type DGs to guarantee Criterion 4.1 effective may be obtained, as shown in Tables 4.12 and 4.13, respectively.

Table 4.8 Load reach limitations to guarantee Criterion 4.1 effective for a radial feeder (km) (Machine-type DGs with $\gamma = 12.5\%$)

Type	Specification	Rated current (A)	DG capacity (MW)	Main side short circuit capacities (MVA)		
				100	200	500
Overhead feeder	LGJ-95	295	1.28	8.1	9.0	9.5
	LGJ-120	335	1.45	7.3	8.4	9.0
	LGJ-150	391	1.69	6.2	7.4	8.0
	LGJ-185	453	1.96	5.2	6.5	7.2
	LGJ-240	536	2.32	4.1	5.4	6.2
	LGJ-300	615	2.66	3.3	4.6	5.4
Cable feeder	YJV-95	298	1.29	19.1	20.4	21.0
	YJV-120	341	1.48	19.4	21.2	22.1
	YJV-150	390	1.69	18.8	21.2	22.4
	YJV-185	444	1.92	17.8	20.8	22.4
	YJV-240	523	2.26	15.7	19.7	21.9
	YJV-300	610	2.64	13.0	17.7	20.2

Table 4.9 Load reach limitations to guarantee Criterion 4.1 effective for a radial feeder (km) (Machine-type DGs with $\gamma = 25\%$)

Type	Specification	Rated current (A)	DG capacity (MW)	Main side short circuit capacities (MVA)		
				100	200	500
Overhead feeder	LGJ-95	295	2.55	3.0	4.0	4.6
	LGJ-120	335	2.90	2.6	3.7	4.3
	LGJ-150	391	3.39	1.9	3.1	3.8
	LGJ-185	453	3.92	1.3	2.6	3.4
	LGJ-240	536	4.64	0.7	2.1	2.8
	LGJ-300	615	5.33	0.2	1.6	2.4
Cable feeder	YJV-95	298	2.58	7.9	9.6	10.3
	YJV-120	341	2.95	7.4	9.7	10.7
	YJV-150	390	3.38	6.4	9.4	10.8
	YJV-185	444	3.85	5.1	8.9	10.7
	YJV-240	523	4.53	3.2	7.9	10.2
	YJV-300	610	5.28	1.1	6.5	9.3

Table 4.10 Load reach limitations to guarantee Criterion 4.1 effective for looped feeders (km) (Machine-type DGs with $\gamma = 12.5\%$)

Type	Specification	Rated current (A)	DG capacity (MW)	Main side short circuit capacities (MVA)		
				100	200	500
Overhead feeder	LGJ-95	295	1.28	3.0	4.0	4.6
	LGJ-120	335	1.45	2.6	3.7	4.3
	LGJ-150	391	1.69	1.9	3.1	3.8
	LGJ-185	453	1.96	1.3	2.6	3.4
	LGJ-240	536	2.32	0.7	2.1	2.8
	LGJ-300	615	2.66	0.2	1.6	2.4
Cable feeder	YJV-95	298	1.29	7.9	9.6	10.3
	YJV-120	341	1.48	7.4	9.7	10.7
	YJV-150	390	1.69	6.4	9.4	10.8
	YJV-185	444	1.92	5.1	8.9	10.7
	YJV-240	523	2.26	3.2	7.9	10.2
	YJV-300	610	2.64	1.1	6.5	9.3

It can be seen from Tables 4.8–4.13 that the adaptability of cables is much higher than that of overhead feeders to guarantee Criterion 4.1 effective. The capacity limitations of machine-type DGs decrease with the increasing load reach.

Table 4.11 Load reach limitations to guarantee Criterion 4.1 effective for looped feeders (km) (Machine-type DGs with $\gamma = 25\%$)

Type	Specification	Rated current (A)	DG capacity (MW)	Main side short circuit capacities (MVA)		
				100	200	500
Overhead feeder	LGJ-95	295	2.55	0.4	1.5	2.1
	LGJ-120	335	2.90	0.0	1.3	1.9
	LGJ-150	391	3.39	—	0.9	1.7
	LGJ-185	453	3.92	—	0.7	1.4
	LGJ-240	536	4.64	—	0.3	1.2
	LGJ-300	615	5.33	—	0.1	1.0
Cable feeder	YJV-95	298	2.58	1.2	4.0	4.9
	YJV-120	341	2.95	—	3.7	5.0
	YJV-150	390	3.38	—	3.2	4.9
	YJV-185	444	3.85	—	2.6	4.8
	YJV-240	523	4.53	—	1.6	4.3
	YJV-300	610	5.28	—	0.6	3.7

Note: the "—" indicates a value is feasible.

Table 4.12 Capacity limitations of machine-type DGs to guarantee Criterion 4.1 effective for cable feeders

Load reach (km)	Main side short circuit capacities (MVA)	Capacity limitations (MW)		
		YJV-70	YJV-150	YJV-240
2	100	4.52/4.24	4.87	4.94
	200	6.79/4.24	8.19/6.75	8.53
	500	8.94/4.24	13.21/6.75	14.80/9.06
4	100	3.40	4.09	4.26
	200	4.28/4.24	6.02	6.61
	500	4.85/4.24	8.03/6.75	9.62/9.06
6	100	2.64	3.49	3.73
	200	3.07	4.69	5.35
	500	3.31	5.73	7.08
8	100	2.14	3.01	3.31
	200	2.38	3.82	4.48
	500	2.51	4.44	5.59
10	100	1.79	2.64	2.96
	200	1.94	3.21	3.85
	500	2.02	3.62	4.62

Note: the h1 in "h1/h2" indicates the theoretical value of the capacity to guarantee Criterion 4.1 effective, h2 indicates the allowed capacity restricted by the rated current of the feeder.

Table 4.13 Capacity limitations of inverter-type DGs to guarantee Criterion 4.1 effective for overhead feeders

Load reach (km)	Main side short circuit capacities (MVA)	Capacity limitations (MW)		
		LGJ-70	LGJ-150	LGJ-240
	100	2.40	2.70	2.77
3	200	2.89	3.47	3.63
	500	3.25	4.16	4.44
	100	1.45	1.74	1.81
6	200	1.59	2.02	2.14
	500	1.69	2.23	2.40
	100	1.03	1.28	1.35
9	200	1.10	1.42	1.52
	500	1.14	1.52	1.64
	100	0.80	1.01	1.07
12	200	0.84	1.09	1.17
	500	0.86	1.15	1.25
	100	0.65	0.83	0.89
15	200	0.68	0.89	0.96
	500	0.69	0.93	1.01
	100	0.55	0.71	0.76
18	200	0.57	0.75	0.81
	500	0.58	0.78	0.84
	100	0.48	0.62	0.66
21	200	0.49	0.65	0.70
	500	0.50	0.67	0.73

Note: the h1 in "h1/h2" indicates the theoretical value of the capacity to guarantee Criterion 4.1 effective, h2 indicates the allowed capacity restricted by the rated current of the feeder.

4.3.3.3.3 Distribution Grids with Hybrid Types of DGs

For the distribution grids with inverter type and machine-type DGs, we should check the adaptability to guarantee Criterion 4.1 effective by short circuit current calculation in extreme fault. If fault current from the main source is always much larger than that from the DGs, we may set an appropriate threshold of I_{set} to report only the fault current from the main source and Criterion 4.1 is effective. Otherwise, Criterion 4.1 is not always effective.

To simplify the problem, we may suppose all the DGs are of the machine type and check the adaptability of Criterion 4.1 using the methodology described for machine-type DGs in the previous subsection. If we find that Criterion 4.1 is effective in that simplified case, it is effectively the cause in the real case. Otherwise, a short circuit current calculation has to be used for further checking.

4.3.3.3.4 An Example

An example is shown in Fig. 4.18. The normal operation mode is shown in Fig. 4.18(a), where S_1 and S_2 are circuit breakers in the substation; A, B, C, E, and F are sectionalizing switches; D is the loop switch; and G and H are the output circuit breakers of DGs. The specification of the feeder is LGJ-240. The numbers in the brackets are the lengths of the corresponding feeder sections. DG1 is a synchronous machine-type DG and DG2 is a doubly fed induction machine-type DG. The rated powers of DG1 and DG2 are both 1.25 MVA. The subtransient reactances of DG1 and DG2 are 0.2 and 0.25, respectively. The main side short circuit capacities of S_1 and S_2 are 500 MVA and 250 MVA, respectively.

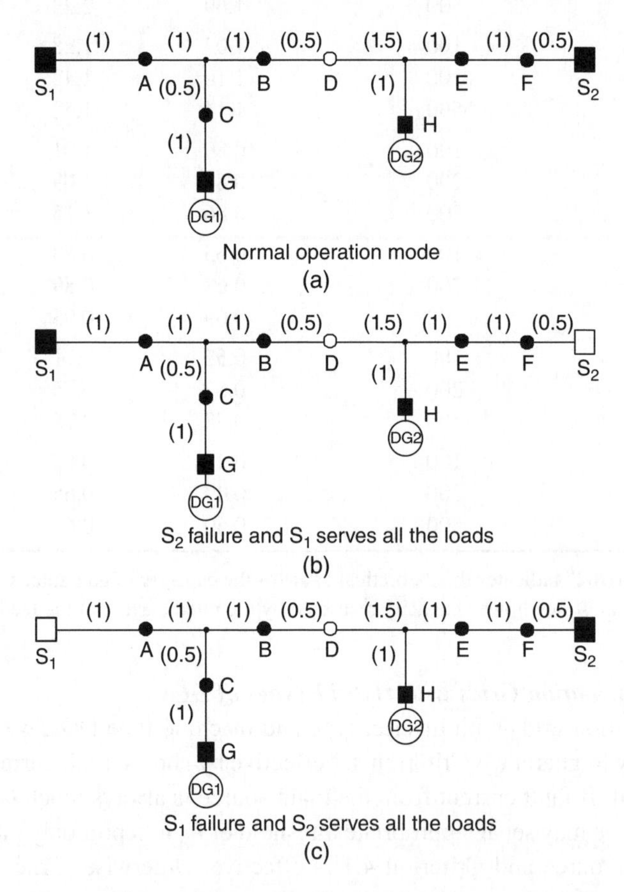

Figure 4.18 An example of looped feeder pair with DGs

Assuming that $\gamma = 12.5\%$, as for S_2, the main side short circuit capacity is 250 MVA but the maximum load reach when the loop switch D is closed is 7.5 km, which exceeds load reach limitation to guarantee Criterion 4.1 effective for overhead feeders according to Table 4.10. Thus, we should check the adaptability to guarantee Criterion 4.1 effective by short circuit current calculation in extreme fault cases.

The case where S_2 fails and S_1 serves all the loads is shown in Fig. 4.18(b). The minimum two phase short circuit current from the main source through the nodes is 1.98 kA while the maximum short circuit current from the DGs is 0.82 kA. The inverse case where S_1 fails and S_2 serves all the loads is shown in Fig. 4.18(c). The minimum two phase short circuit current from the main source through the nodes is 1.85 kA while the maximum short circuit current from the DGs is 0.76 kA.

From this calculation, as for the distribution grid in Fig. 4.18, the maximum short circuit current from the DGs is 0.82 kA and the minimum two phase short circuit current from the main source is 1.85 kA. We may set a threshold of 1.42 kA to guarantee Criterion 4.1 effective.

It can be seen from the example that even if the limitations in Tables 4.2–4.13 are not satisfied, we still have the possibility to be able to find an appropriate threshold to guarantee Criterion 4.1 effective, because the limitations in Tables 4.2–4.13 are worked out for the worst conditions, which are far stricter than conditions in practice. In other words, once the limitations in Tables 4.2–4.13 are satisfied, Criterion 4.1 can be guaranteed effective with sufficient confidence.

4.3.3.4 Methodologies of Fault Location for DGs with Larger Capacities

It can be seen from Tables 4.2–4.13 that wide scopes of traditional fault location approaches are effective in locating fault position for distribution grids with low capacity DGs. In this subsection, we describe some methodologies of fault location for DGs with larger capacities.

4.3.3.4.1 Fault Location based on the Coordination of the Reclosing Procedure and Cutting Out DGs

After a fault occurs, a certain circuit breaker is tripped by the corresponding relay protection device causing the feeder outage. The DGs attached to the feeder cut out via the anti-islanding protection within 2 sec. The fault location methodology based on this feature is coordinated with the reclosing procedure; the main measures are as follows:

- The time delay of reclosing is changed to 3–4 sec. for circuit breakers.
- In case a fault occurs, a certain circuit breaker trips with its reclosing delay timer started.
- 2 sec. after the fault, all DGs on the feeder with fault are cut out by the anti-islanding protection.
- 3–4 sec. after the fault, the tripped circuit breaker recloses. If it is a temporary fault, the service is restored and the DGs will connect to the grid quickly. If it is permanent, the fault phenomenon is the same as the feeder without DGs where the short circuit current comes only from the main source point, therefore the fault can be located according to traditional fault location approaches no matter how large the capacities of the DGs.

Notably, as for a permanent fault, the short circuit current from the main source node when reclosing includes not only the short circuit current from the main power source but also that from the DGs attached to the other feeders connected to the bus serving the fault feeder.

As for the example shown in Fig. 4.19, S_1 is the main source node, that is, the circuit breaker in the substation; A, B, C, D, and E are sectionalizing switches; F and H are the output circuit breakers of DGs; DG1 is a photovoltaic power source with large capacity; and DG2 is a micro gas turbine with large capacity. Assume that the traditional fault location approaches are not effective in this example.

First, we investigate the case where a temporary fault occurs in the region of Re(B,D). When the fault happens, S_1 is quickly tripped to clear the fault as shown in Fig. 4.19(a). After 2 sec., F and H open cutting out the DGs as shown in Fig. 4.19(b). One more seconds later, S_1 recloses and succeeds as shown in Fig. 4.19(c). In a short time, F and H are closed connecting the DGs to the feeder as shown in Fig. 4.19(d). The distribution grid restores to its normal state.

Next, we investigate the case where a permanent fault occurs in the region of Re(B,D). When the fault happens, S_1 is quickly tripped to clear the fault as shown in Fig. 4.19(e). S_1, A, B, D, E, F, and H all send the fault information to the master station of the DAS, indicated by arrows. Thus the fault cannot be located according to the traditional fault location approaches with only fault current information and without the direction information of the fault power. After 2 sec., F and H are open cutting out the DGs. One more second later, S_1 recloses and trips again. This time, only S_1, A, and B send the fault information to the DAS master station. The fault can be correctly located according to the traditional fault location approach as shown in Fig. 4.19(f).

4.3.3.4.2 Fault Location based on an Over-Heat Region Searching Approach

As discussed in the preceding subsection, Criterion 4.2 is effective at locating faults on distribution grids with DGs. Criterion 4.2 is based on the information about fault power directions. But in practice, the directions of fault power from DGs may be wrong due to the fault current being too small to be accurately detected. Sometimes, even a DG with large capacity may provide a small fault current due to lack of primary resources, such as sunshine, wind, and so on. Once the wrong information about the direction of fault power exists, the fault cannot be correctly located according to Criterion 4.2.

To solve that problem, fault location based on the over-heat region searching approach is introduced.

For a region, we may define a sensitivity parameter K_{sen}, if Equation (4.34) is satisfied, we call the region an *over-heat region*.

$$L_R > K_{sen} L_{R,\max} \tag{4.34}$$

Where, $L_{R,\max}$ is the load limitation of the region, L_R is the load distributed from the region.

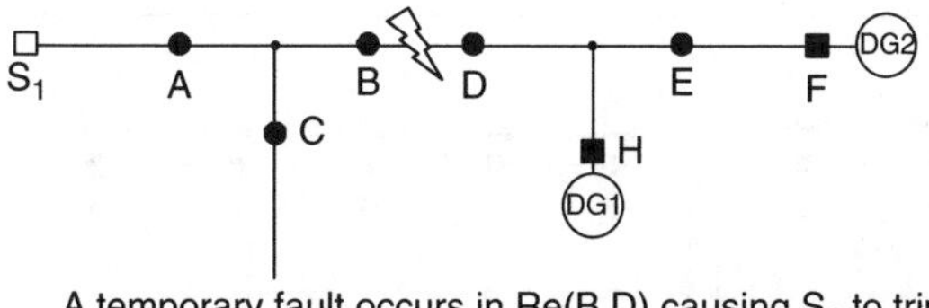

A temporary fault occurs in Re(B,D) causing S_1 to trip

(a)

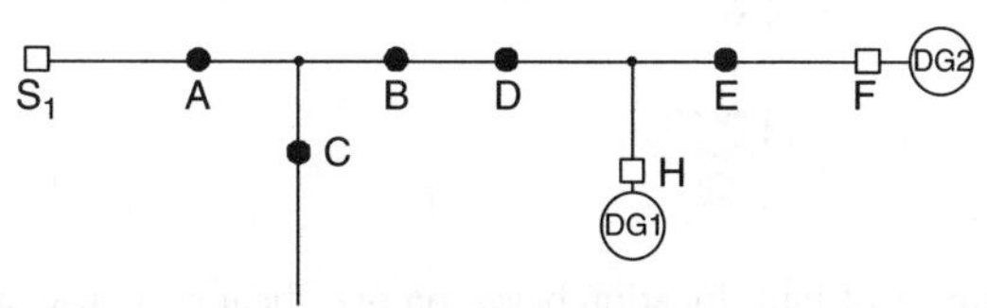

DG1 and DG2 cut out after 2 sec.

(b)

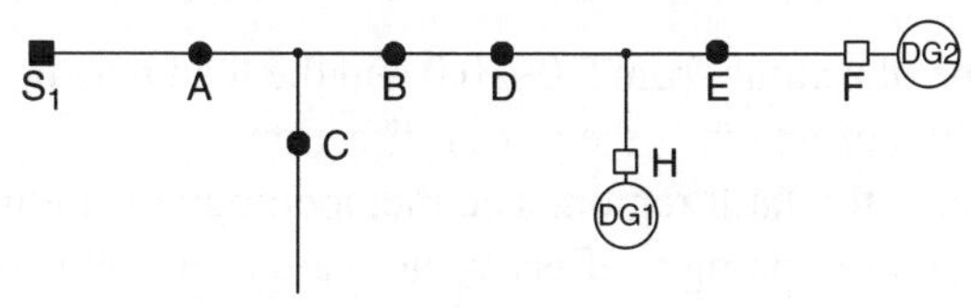

S_1 recloses and succeeds 3 sec. after the fault

(c)

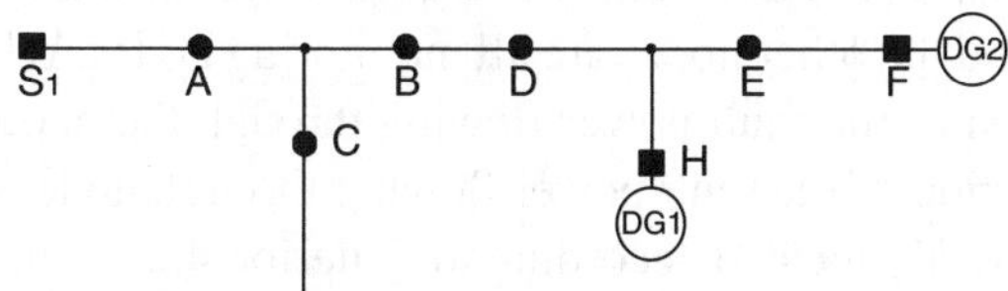

DG1 and DG2 connected to the feeder. The temporary fault processing is finished

(d)

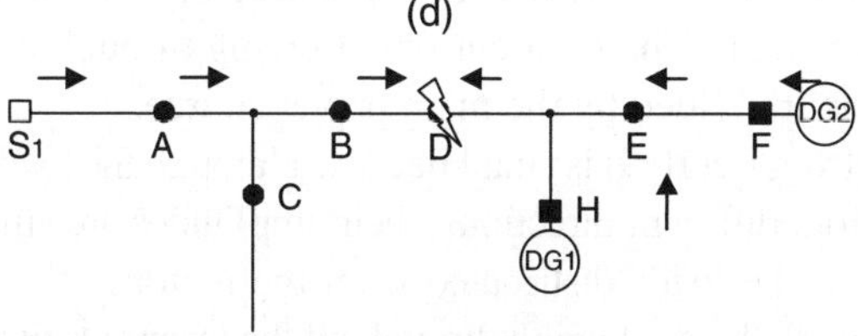

A temporary fault occurs in Re (B,D) causing S_1 to trip. S_1, A, B, D, E, F, and H all send the fault information,indicated by arrows. The traditional fault location approaches based on fault current are not effective.

(e)

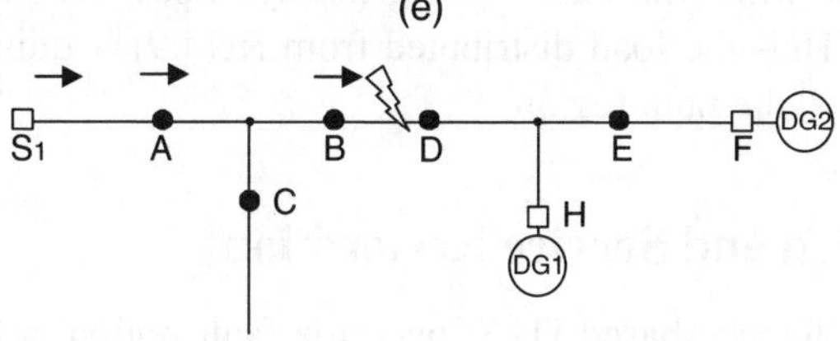

Two seconds after the fault, DG1 and DG2 cut out. One more second later, S1 recloses and trips again. Only S1, A, and B send the fault information. The fault is located according to Criterion 4.1.

(f)

Figure 4.19 An example of fault location based on coordinating the reclosing procedure and cutting out the DGs

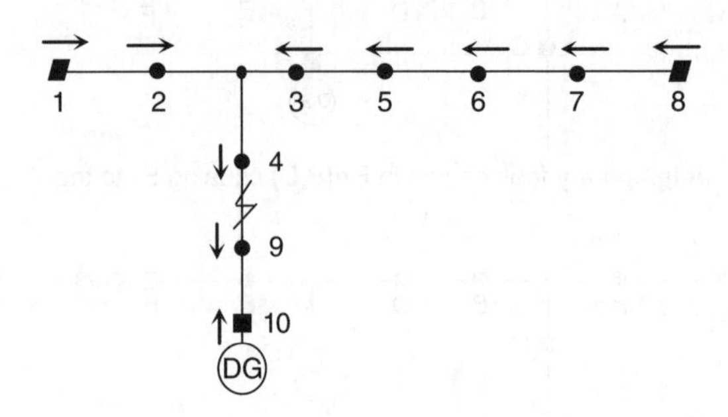

Figure 4.20 An example of fault location based on over-heat region searching approach

Generally, K_{sen} can be of a value from 5.0~10.0 and the load may be indicated by the load current, the direction of which is the power direction.

The over-heat region is the fault region. The methodology of finding fault location based on over-heat region searching is effective no matter how large or how small the capacities of DGs.

Look at the example in Fig. 4.20. Node-1 and node-8 are the main sources, node-10 is a DG source node, and the other nodes are all monitoring nodes. When a fault occurs in Re(4,9), the direction of the fault power flowing through the nodes is indicated by arrows. Note, the direction of the fault power flowing through node-9 is wrong. Thus, the fault cannot be correctly located according to Criterion 4.2.

The reason for this wrong information about power through node-9 is the fault current through there is too small to be accurately detected. The fault current flowing through node-10 is also small. The fault currents flowing through the other nodes are quite large since they are provided by the main power source.

The load distributed from Re(10,9) is small because although the loads flowing through both its terminals are from different directions, their amplitudes are small, so Re(10,9) is not an over-heat region. The loads distributed from the regions other than Re(4,9) are small due to the fact that the loads flowing through all the terminals of the corresponding regions are large and show little difference. Therefore, they are not over-heat regions.

For the region Re(4,9), the load flowing through node-4 is quite large since it is from the main power source while the load flowing through node-9 is very small since it comes from the DG. Thus the load distributed from Re(4,9) is quite large and is an over-heat region, that is, the fault region.

4.4 Fault Isolation and Service Restoration

For a centralized intelligence-based DAS, once the fault region is located, the fault isolation program should be executed by tele-control operation, after that the service restoration program should be carried out.

4.4.1 Fault Isolation

It was pointed out in Section 4.2 that a monitoring region with terminals of monitoring nodes is the smallest unit of fault location, while an action region with terminals of action nodes is the smallest unit of fault isolation and load transfer. In the practice, the scope of a monitoring region is usually different to the corresponding action region because a monitoring node with a fault indicator does not have the ability to be remote controlled by the DAS master station. But an action node with FTU generally has the function of fault information reporting, which is also often a monitoring node. Thus, the scope of an action region always embodies the corresponding monitoring region. In other words, the scope of an action region is always larger than the corresponding monitoring region.

Therefore, the first thing to do for fault isolation is to *find out the corresponding action region for the located fault region*. A feasible procedure for this, which is called FAR, is as follows:

Step 1: Find a terminal of the located fault region to deal with.
Step 2: If the terminal is an action node, put it into Set $\boldsymbol{Q}$, go to Step 4.
Step 3: If the terminal is an ending node, put it into Set $\boldsymbol{Q}$. Otherwise, find the adjacent action nodes and ending nodes in the direction from the inside of the region to the outside, put the action nodes into Set $\boldsymbol{Q}$.
Step 4: If all of the terminals are handled, the process ends, and the corresponding action region is the region with the terminals in Set $\boldsymbol{Q}$. Otherwise, go to back to Step 1.

The second thing to do for fault isolation is carry out the *remote control process of fault isolation* to trip the terminals, which are in closed states, of the corresponding action region to the located fault region by remote control from the DAS master station. Notably, some of the terminals have been tripped before the remote tripping process by relay protection devices or other mechanisms, for example tripping due to loss of voltage mechanism, and so on. Some of the terminals are open, such as loop switches, the ending nodes, and so on. Thus, only the terminals still in closed states should be tripped by remote control.

If all tripping controls are successful, the fault isolation program ends. Otherwise, we should carry out the *re-tripping process* to re-trip the switches that failed to trip three times; once a tripping control is successful, the re-tripping process of the corresponding switch ends.

If all of re-tripping processes are successful, the fault isolation program ends. Otherwise, we should carry out the *correction control process of fault isolation* as follows:

Step 1: Change the node type that has failed to trip from action to monitoring node type.
Step 2: Divide the monitoring regions and action regions according to the new node types.

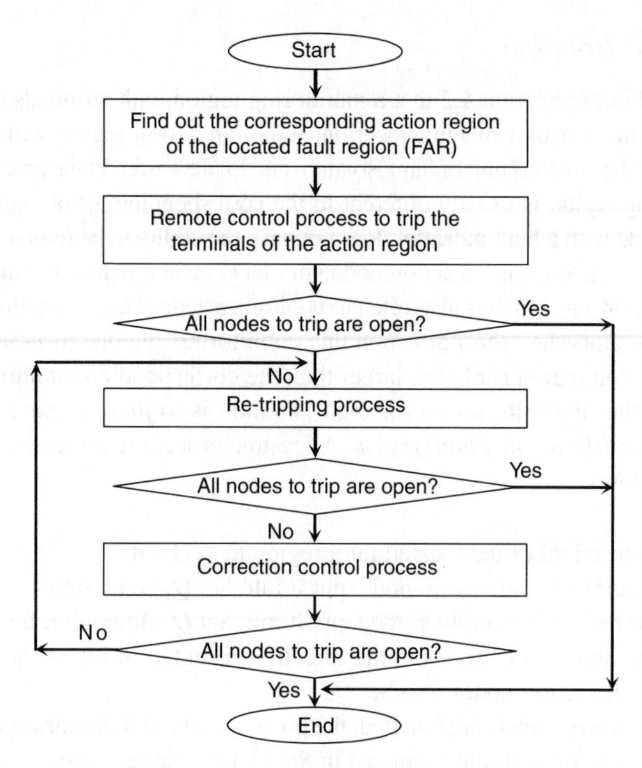

Figure 4.21 The whole program of fault isolation

Step 3: Find out the new located fault region, which is the smallest monitoring region containing the previously located fault region.

Step 4: Find out the new action region according to the new located fault region by carrying out the steps of FAR.

Step 5: Carry out the remote tripping process.

The whole program of fault isolation is shown in Figure 4.21. Note in particular that this program of fault isolation will completely end the process because, as the action region changes become larger and larger due to some nodes failing to trip by remote control, the terminals of the new action region will finally reach the source nodes, the loop switches, or the ending nodes, which are all in open states. Thus, the final condition of "All nodes to trip are open?" will be finally satisfied.

As for the examples in Figs 4.2–4.4, supposing that the located fault region is Re(B_3, D_3). Since its terminal node-B_3 is not an action node, we should find out its adjacent action nodes in the directions from inside the region to the outside (i.e., node-A_3, node-A_4 and node-B_4) and put them into Set $\boldsymbol{Q}$. Since node-D_3 is an ending node, it is also put into Set $\boldsymbol{Q}$. Thus, $\boldsymbol{Q}$ = [A_3, A_4, D_3, B_4]. The corresponding action region is Re(A_3, A_4, D_3, B_4).

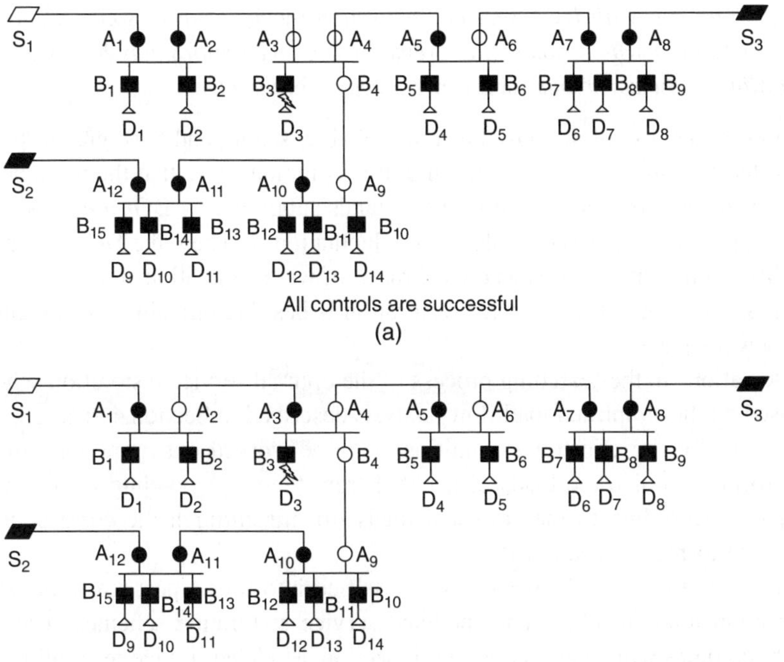

Figure 4.22 Examples of fault isolation for the distribution grid in Fig. 4.2

Carrying out the remote control fault isolation process to trip the terminals, if all controls are successful, fault isolation is completed as shown in Fig. 4.22(a).

If the remote control of A_3 tripping failed and the re-tripping process of A_3 also failed, the correction control process will trip A_2 by remote control. Supposing that the correction control of A_2 is successful, the fault isolation is completed as shown in Fig. 4.22(b).

4.4.2 *Service Restoration*

For a centralized intelligence based DAS, service restoration can be arranged in an optimized program, which can be designed as an optimization problem and will be discussed in this section.

The service restoration program may be carried out within the connected systems to the fault regions, respectively.

Criterion 4.6 is suggested for the service restoration of the distribution grids with DGs, that is,

For distribution grids with DGs, service restoration by main power sources is of the highest priority, after which, some of the remained outage regions may be restored by DGs according to the conscious islands arranged beforehand.

Main power sources are not only far more reliable, stable, and have greater power capacities, but they are also more economic. It is common sense that the more power capacity a grid has, the more safe it is. The islands supported by DGs are sometimes not stable, especially in cases of violent load fluctuation or where the DGs lack regulating ability, such as in the cases where the ratios of wind generation and photovoltaic generation are too high. Thus, service restoration is best being mainly dependent on the main power source.

The calculations in the searching process of the optimal service restoration scheme can be based on the simplified load flow analysis described in Section 4.2; whether or not object functions or constraint conditions may be worked out by region-to-node load transform, in which the loads distributed from the corresponding regions have been prepared according to the collected loads flowing through the corresponding nodes by node-to-region load transform.

The service restoration schemes can be classified into four types; real time, short time, minimum load shedding, and modeled service restoration schemes. The first three methodologies will be discussed in this section. Modeled service restoration will be described in Section 4.4.3.

4.4.2.1 Real Time Service Restoration

For *real time service restoration*, its control scheme is produced based on the loads of the regions collected just before the moment of fault occurs. No load shedding due to overload is needed in the real time service restoration program, otherwise it cannot be considered as a real time service restoration scheme.

4.4.2.1.1 Objective Function

Usually, the more balanced the loads distributed in the grid, the higher a loading ability it has. Thus, the index to indicate how balanced the loads are in a connected system (i.e., Load Balance Degree: LBD), which is defined as Equation (4.35), may be used as the objective function shown in Equation (4.36).

$$LBD = \frac{I'_{N,\min}}{I'_{N,\max}} \tag{4.35}$$

Where, $I'_{N,\min}$ and $I'_{N,\max}$ are the minimum and maximum relative loading through the main source nodes in the close state in the connected system with the fault region. Of course, the larger the LBD is, the more balanced the loads in the connected system.

$$\max \quad f_1 = \frac{I'_{N,\min}}{I'_{N,\max}} \tag{4.36}$$

Sometimes, to minimize the number of nodes to control, the following may also be used as the objective function shown in Equation (4.37).

$$\min \quad f_2 = \sum_{k \in S_s} (1 - K_k) + \sum_{k \in T_s} K_k \tag{4.37}$$

Where, $\boldsymbol{T}_{\mathrm{S}}$ and $\boldsymbol{S}_{\mathrm{S}}$ are the sets of the normally open and normally closed action nodes before the fault occurred. K_k indicates the state of the k-th action node. $K_k = 1$ and 0 and indicates the k-th action node being closed and open, respectively.

As for the conditions of service restoration, restoring as much load as possible is important. Thus, economic operation indexes are seldom used as objective functions.

4.4.2.1.2 Constraint Conditions

Usually, the current limitations of the distribution facilities are used as the constraint conditions shown in Equation (4.38).

$$I'_{N,\max} < 1.0 \tag{4.38}$$

The facility flowing through the largest current of a feeder is the main source node. Thus, generally, only the main source nodes should be checked according to Equation (4.38). It is worth mentioning that $I'_{N,\max}$ is a value taking the decrease of DGs into consideration and the re-connection of the DGs after service restoration by the main power source needs a period of time, during which, the actual value of $I'_{N,\max}$ is a little larger, sometimes even larger than 1.0. But such a situation may only last a short period and can be endured without any damage.

The topology constraints are also used as constraint conditions for real time service restoration, which consists of two aspects; no loops among main source nodes exist and no more outage islands occur than *the simplest restoration condition.*

The so called simplest restoration condition is where the fault region is isolated and all restorable regions are restored by closing the source node or the action nodes of looped switches regardless of whether the current limitations are satisfied, but the topology constraint of no loops among main source nodes must be satisfied.

The control constrains are also used as constraint conditions, according to which, some action nodes are not allowed to control due to damage, maintenance, and other reasons.

As a distribution grid is carefully designed considering contingency and emergency conditions, in which voltage limitations may not be as strict as normal conditions, the voltage limitations are seldom meant to be used as constraint conditions.

4.4.2.1.3 Solution Space

A service restoration scheme can be regarded as a candidate solution, which is formed of a set of node states. If the i-th element is "1", it indicates the i-th action node is in the closed state. If the i-th element is "0", it indicates the i-th action node is in the open state. Thus, the solution can be coded in binary.

According to the topology constraints, the number of action nodes in open states is constant for each feasible restoration scheme. Therefore, the solution can also be formed of a set of sequence numbers of only the nodes in an open state.

4.4.2.1.4 Searching Approach

The service restoration optimization is a NP-hard problem: the searching approach to find out the optimized solution is rather complicated. Many achievements have been made in this area, such as the Gene Algorithm (GA), Particle Swarm Optimization algorithm (PSO), the branch exchange algorithms, Optimal Flow Pattern algorithm (OFP), the Simulated Annealing algorithm (SA), Artificial Neural Network algorithm (ANN), Expert System based approach (ES), Fish Swarm algorithm (FS), Fuzzy Evaluation approach (FE), Tabu Search algorithm (TS), and so on.

The Tabu Search algorithm (TS) is used as an example detailed here.

A candidate solution is formed of a set of the sequence numbers of only the nodes in an open state. As for the k-th iteration, the current solution $\boldsymbol{sw}^{(k)}$ is expressed as

$$\boldsymbol{sw}^{(k)} = \left[sw_1^{(k)} \quad sw_2^{(k)} \quad \ldots \quad sw_M^{(k)} \right] \tag{4.39}$$

Where, $sw_i^{(k)}$ is the sequence number of the i-th open-state node, M is the number of the elements in $\boldsymbol{sw}$.

Equation (4.36) is used as the adaptive function. The constraint conditions include Equation (4.38) and the topology constraints.

In the Tabu searching approach, the position of each open-state node in the current solution is changed to its adjacent action node in the direction of the upstream and downstream, respectively, thus forming the set of candidate solutions. As for the k-th iteration, the set of new solutions is

$$\boldsymbol{CW}^{(k)} = \left[\boldsymbol{cw}_1^{(k)} \quad \boldsymbol{cw}_2^{(k)} \quad \ldots \quad \boldsymbol{cw}_{2M}^{(k)} \right],$$

in which,

$$\boldsymbol{cw}_i^{(k)} = \left[sw_1^{(k)} \quad sw_2^{(k)} \quad \ldots \quad sw_i^{(k)+} \quad \ldots \quad sw_M^{(k)} \right] \quad (\text{for } i = 1 \sim M) \tag{4.40}$$

$$\boldsymbol{cw}_i^{(k)} = \left[sw_{M+1}^{(k)} \quad sw_{M+2}^{(k)} \quad \ldots \quad sw_i^{(k)-} \quad \ldots \quad sw_{2M}^{(k)} \right] \quad (\text{for } i = M \sim 2M) \tag{4.41}$$

Where, $sw_i^{(k)+}$ and $sw_i^{(k)-}$ indicate the upstream and downstream adjacent action nodes of a current open-state node $sw_i^{(k)}$, respectively.

The set of candidate solutions (i.e., $\boldsymbol{DW}^{(k)}$) is formed by the elements satisfying the constraint conditions in $\boldsymbol{CW}^{(k)}$.

The adaptive function of Equation (4.36) is calculated for each candidate solution in $\boldsymbol{DW}^{(k)}$.

If the value of the adaptive function of the best candidate solution in $\boldsymbol{DW}^{(k)}$ is superior to the "best so far" solution, its Tabu count is ignored and it now becomes the best so far solution instead of the old one. Also, the sequence number of the node that has newly changed from the closed to open state should be put into the Tabu table and the counters of the nodes in the Tabu table should be renewed. If this superior solution does not exist, the candidate solution with the highest adaptive function value in $\boldsymbol{DW}^{(k)}$ is chosen to be the current solution of the $(k+1)$-th iteration (i.e., $\boldsymbol{sw}^{(k+1)}$), no matter whether it is superior to $\boldsymbol{sw}^{(k)}$. The sequence number of the newly changed node from closed to open state should be put into the Tabu table and the counters of the nodes in the Tabu table should be renewed.

The end condition is often the desired iteration times set beforehand.

A general process of an optimal real-time service restoration scheme searching is shown in Fig. 4.23.

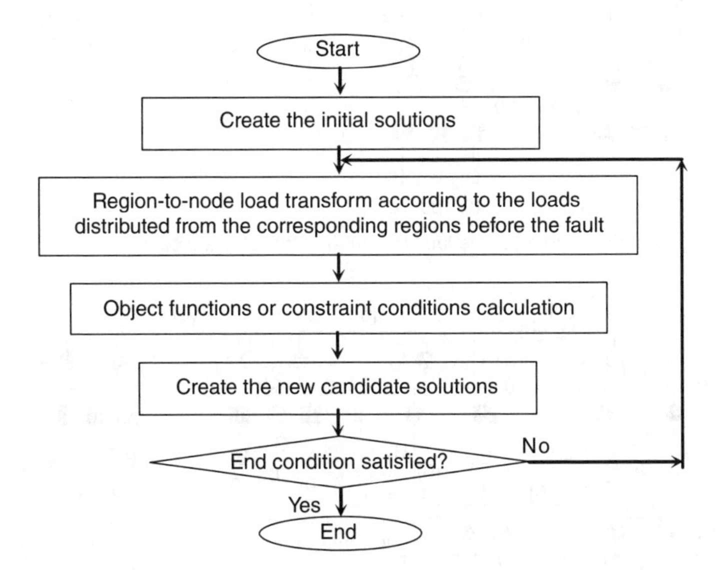

Figure 4.23 A general process of optimal real-time service restoration scheme searching

4.4.2.1.5 An Example

For the connected system shown in Fig. 4.24(a) the rated current of all of the nodes is 400 A. The data beside the nodes indicate the loads (A) flowing through the nodes while the data in brackets "( )" indicate the loads (A) distributed from the action regions and the data in square brackets "[ ]" indicate the loads (A) distributed from the action regions just before the fault. S_1, S_2, S_3, and the nodes indicated by circles are

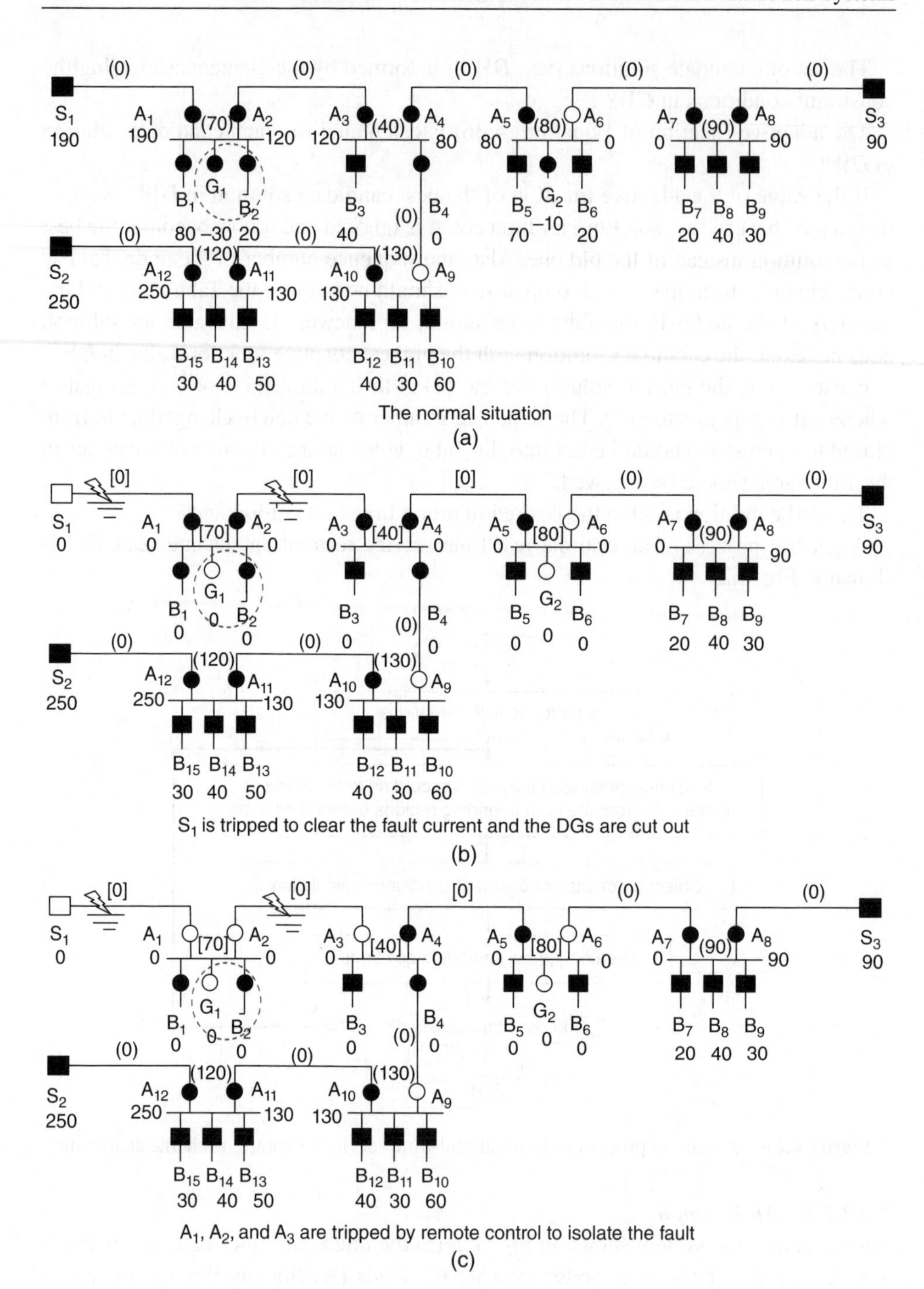

Figure 4.24 An example of real-time service restoration

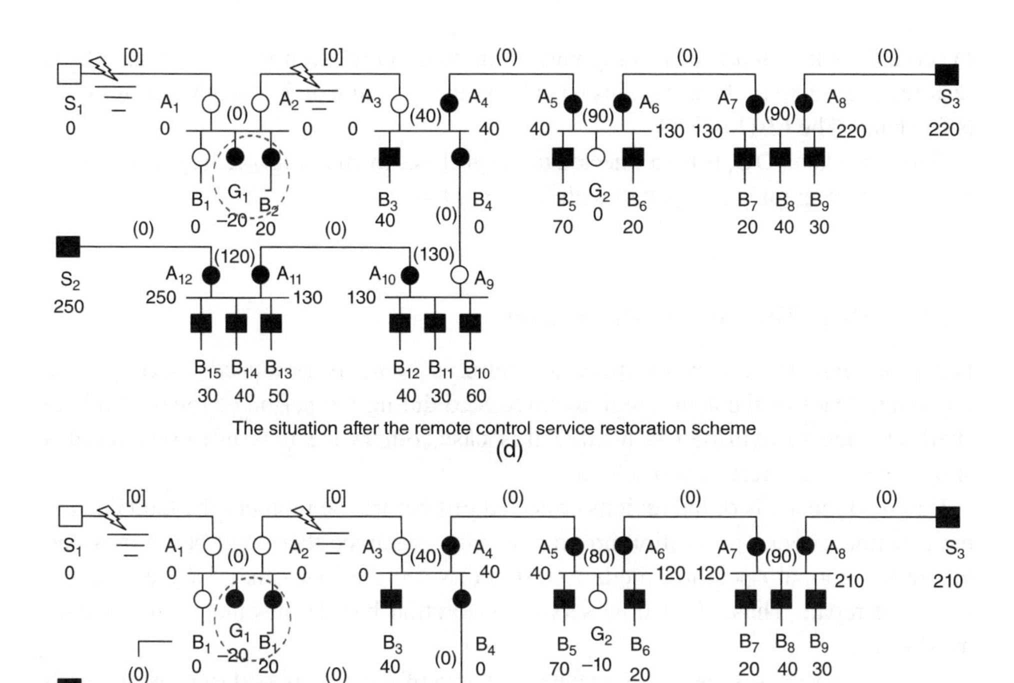

The situation after the remote control service restoration scheme

(d)

DG$_1$ is re-connected to the grid, the service restoration program is completed

(e)

Figure 4.24 (*continued*)

action nodes. S_1, S_2, and S_3 are main source nodes, G_1 and G_2 are DG source nodes. The only conscious island is indicated by the dashed circle shown in the figure containing G_2 and B_2. Because the power capacity of G_1 is too small, it cannot support any island.

Supposing a permanent fault of two phase short circuit to ground occurs, in which the grounding position of phase-A is in Re(A_2, A_3) and the grounding position of phase-C is in Re(S_1, A_1), S_1 is tripped to clear the fault current then G_1 and G_2 are cut out by the anti-islanding protection within 2 sec., which is shown in Fig. 4.24(b). It can be seen from the figure that the loading of S_2 is much heavier than S_3. The LBD goes as low as 0.36.

Since the other terminals of the fault region – that is, A_1, A_2, and A_3 – are all action nodes, they are tripped by remote control to isolate the fault shown in Fig. 4.24(c).

The situation after the remote control of service restoration scheme is shown in Fig. 4.24(d). It is worth noting that the loads between node-A_2 and the source node-S_1 would not be served without DG$_2$. But the output power capacity of DG$_2$ is too small to serve the whole load. Thus, only part of the loads between node-A_2 and node-S_1 is restored according to the conscious island arrangement in which enough margin is left beforehand. The LBD is 0.88.

Sometime later, DG_1 is re-connected to the grid as shown in Fig. 4.24(e). The service restoration program is completed with LBD = 0.84.

4.4.2.2 Short-Time Service Restoration

For *short-time service restoration*, a control scheme is produced based on the maximum loads of the action regions forecasted during the period of repair. No load shedding due to overload is needed this case, otherwise it is not considered a short-time service restoration scheme.

Based on the loads of the regions collected just before the moment the fault occurs, the real time service restoration program cannot guarantee there will be no overload before the normal operation mode is restored, since the loads may vary during the period of repair. Thus, short-time service restoration has the advantage of avoiding overloading.

Many achievements have been made in the area of short term load forecasting, such as the extrapolation method, wavelet transform based approach, Artificial Neural Network (ANN) based approach, Support Vector Machine (SVM) based approach, gray prediction based method, and so on. In practice, the forecast results, even just by the statistical methods based on the history loads of recent similar days, may be satisfied for ultra-short term load forecasting.

It is worth mentioning that the behavior of a single load is random and cannot be satisfactorily forecasted but a certain regularity is shown by a group of loads. The larger the number of loads in a group, the stronger the regularity shown. Thus, we suggest that load forecasting should not be carried out for very small regions. In fact, the number of action nodes is not too many in practice so, an action region may usually contain a reasonable number of loads.

The objective function, constraint conditions, solution space, and searching approaches of short time service restoration are the same as that for real time service restoration. The former is far stricter than the latter. In other words, we may find a real time service restoration scheme even if no feasible short time service restoration schemes exist.

4.4.2.3 Service Restoration with Minimum Load Shedding

Load shedding due to overload is not permitted for either short time service restoration or real-time service restoration. In some cases, such as bus voltage loss due to fault on the buses, transformers, or transmission lines, we need to shed some loads to restore the service as much as possible. This job is realized by the *service restoration with minimum load shedding program.*

This often involves many related connected systems.

4.4.2.3.1 Methodology

The objective functions for service restoration with minimum load shedding may take the following forms shown in Equations (4.42) and (4.43).

The objective function of minimum load shedding is

$$\min \quad f_1 = \sum_{k \in \beta} L_{R,k} \tag{4.42}$$

Where, β is the set of load shedding regions.

The objective function of maximum load restored with fewer switching operations, that is,

$$Max \quad f_2 = \frac{\sum_{i \in \gamma} k_i L_{R,i}}{1 + \alpha T} \tag{4.43}$$

Where, γ is the set of the regions being restored, k_i is the parameter indicating the economic losses of the load shedding of the *i*-th region, *T* is the number of switches needed to remote control, and α is the weighted parameter.

The current limitations of the distribution facilities and the control constrains are also used as the constraint conditions, which are the same as the constraint conditions for short- and real-time service restoration. But the topology constraint only consists of one aspect, that is, no loops among main source nodes exist, which is different from that of short time and real time service restoration.

The solution space and searching approach of the service restoration with minimum load shedding are similar to that of short- and real-time service restoration.

4.4.2.3.2 Approach based on Tabu Search

In this subsection, an approach based on Tabu searching is used as an example for service restoration with minimum load shedding.

The Tabu searching approach for real- and short-time service restoration cannot support load shedding due to overload, because the number of the action nodes in open states is constant for each feasible restoration scheme.

To solve this problem, a concept of the *virtual open action node* (VOAN) is introduced. The number of VOANs is the sum of the ending node amount and the main source node amount. The initial positions of the VOANs are at the positions of the ending nodes and beside the main source nodes, which are shown in Figure 4.25.

With the help of virtual open action nodes (VOAN), various operation modes can be produced in the Tabu search. In the example shown in Fig. 4.25, the candidate

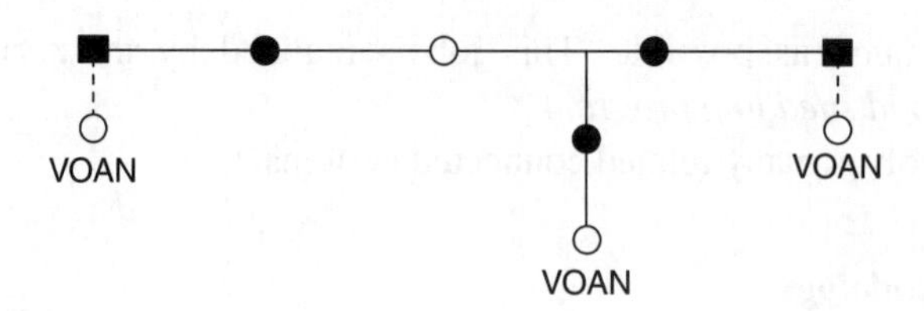

Figure 4.25 Initial positions of virtual open action nodes (VOAN)

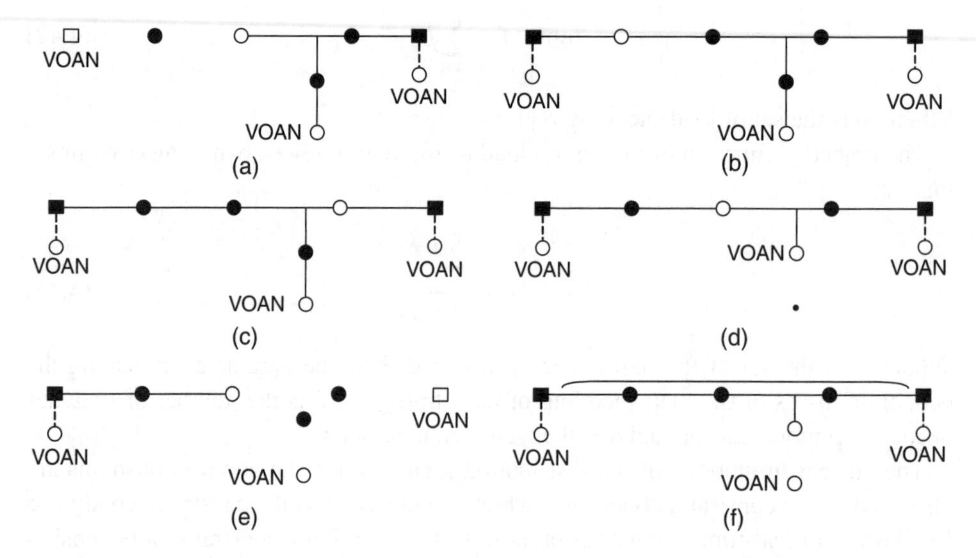

Figure 4.26 Candidate solutions in Tabu searching for the example shown in Fig. 4.25

solutions are shown in Fig. 4.26(a)–(e), respectively. The load shedding is shown in the corresponding figure.

The case of all the VOANs in their initial positions, that is, at the positions of the ending nodes and beside the main source nodes, means that all loads are restored without any load shedding, which are shown in Fig. 4.26(b) and (c). Various operation modes with load shedding can be produced, which are shown in Fig. 4.26(a), (d), and (e). The operation mode in Fig. 4.26(f) is not feasible, since there is a loop between two main source nodes.

As for the example in Fig. 4.27(a), Equation (4.42) is used as the object function. Supposing that a permanent fault occurs in Re(A_2, A_3), S_1 is tripped to clear the fault current, which is shown in Fig. 4.27(b). Since S_3 and S_2 are all heavy loading, load shedding is needed. The operation mode after service restoration with minimum load shedding is shown in Fig. 4.27(c). The load shedding is 120 A. The LBD is 0.81.

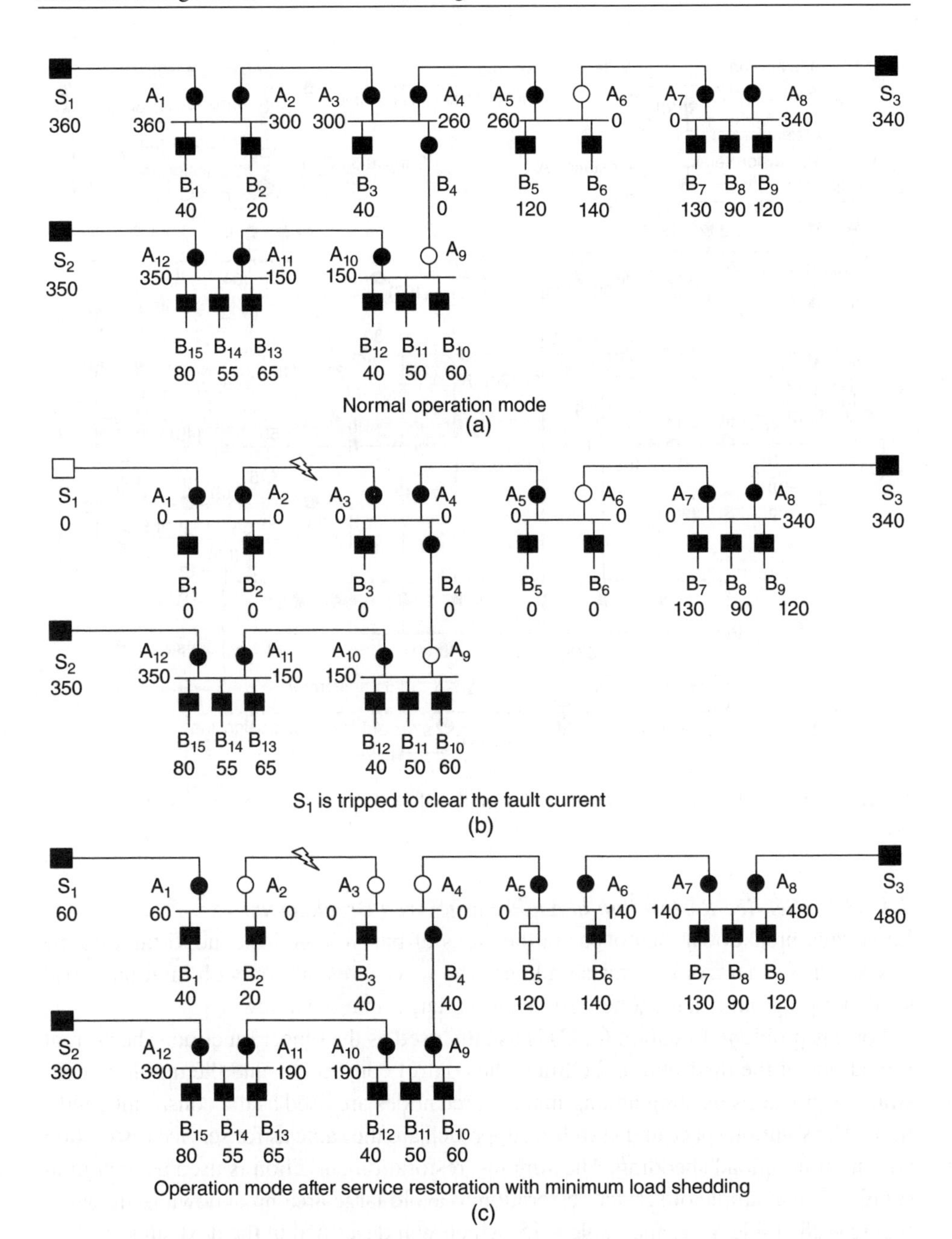

Figure 4.27 An example of service restoration with minimum load shedding

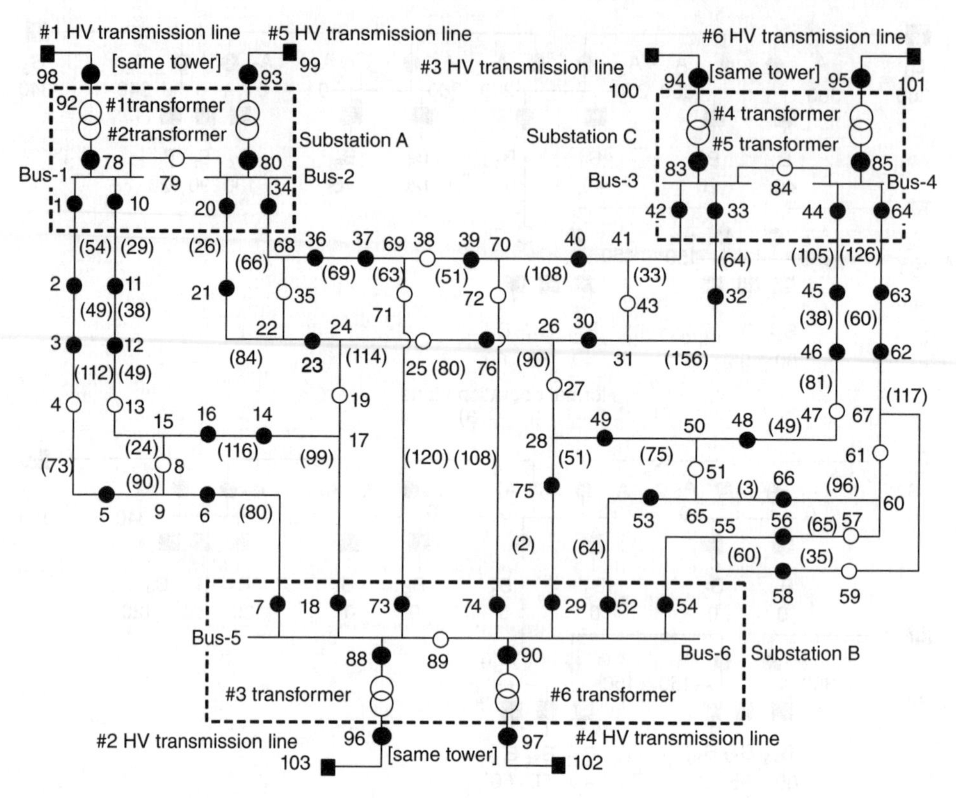

Figure 4.28 A distribution grid as an example of service restoration to avoid large area breakdown

4.4.2.3.3 Service Restoration to Avoid Large Area Breakdown

Large area breakdown may occur in the cases of bus voltage loss due to fault on the buses, transformers, or transmission lines. The service restoration with minimum load shedding program is a great help for avoiding large area breakdown.

For this problem, Equation (4.43) is usually used as the object function. The current limitations of the distribution facilities, the control constraints, and the topology constraint of no loops existing among main source nodes are used as the constraint conditions. The solution space and searching approach are the same as for service restoration with minimum load shedding. The simplest restoration condition is used as the initial solution. An example of service restoration to avoid large area breakdown is shown in Figure 4.28, Table 4.14, and Table 4.15, which will described in the next subsection.

4.4.2.4 Remote Control Steps of Service Restoration

In the previous subsections, the methodology to obtain service restoration schemes is dealt with. But in practice, we should arrange remote control steps from current

Table 4.14 The outage losses of the action regions of the example shown in Figure 4.28

Action regions	Outage losses ($)	Action regions	Outage losses ($)
Re(10,11)	3293.95	Re(14,16)	62,925.64
Re(2,3)	5566.49	Re(4,5)	39,599.75
Re(11,12)	16,559.30	Re(5,6,8)	48,821.61
Re(37,38,71)	7155.82	Re(71,73)	65,095.49
Re(38,39)	5792.81	Re(21,23,35)	45,566.84
Re(33,32)	7269.41	Re(27,30,76)	48,821.61
Re(29,75)	227.17	Re(30,32,43)	84,624.13
Re(25,76)	9086.76	Re(48,49,51)	40,684.68
Re(14,18,19)	43,141.35	Re(56,57)	35,260.05
Re(1,2)	4595.96	Re(47,48)	26,580.65
Re(8,13,16)	2042.65	Re(45,46)	20,613.57
Re(20,21)	2212.87	Re(63,64)	68,350.26
Re(34,35,36)	5617.29	Re(36,37)	30,068.21
Re(40,42,43)	2808.64	Re(19,23,25)	49,677.91
Re(27,49,75)	4340.63	Re(39,40,72)	47,063.29
Re(52,53)	5447.07	Re(72,74)	47,063.29
Re(54,56,58)	5106.63	Re(59,61,62)	50,985.23
Re(58,59)	2978.86	Re(57,61,66)	41,834.03
Re(62,63)	5106.63	Re(44,45)	45,755.97
Re(51,53,66)	255.33	Re(46,47)	35,297.46
Re(6,7)	43,396.99	Re(3,4)	48,806.37
Re(12,13)	21,352.79		

Note: The outage losses here apply to an outage of 4 h.

operation mode to the operation mode desired by service restoration scheme, especially for service restoration to avoid a large area breakdown, in which many switches should be tele-controlled. If the remote control steps are not arranged well, overload may occur during the tele-control procedure.

4.4.2.4.1 Some Operation Modes

Some operation modes related to fault processing are as follows:

The *normal operation mode* is the regular operation mode of a distribution grid.

The *fault isolation operation mode* is the operation mode with the faults having been isolated.

The *service restoration operation mode* is the operation mode with the service restoration scheme having been carried out.

The control steps from normal operation mode to fault isolation operation mode are not discussed since they are rather simple with the only controls to open the corresponding switches.

Table 4.15 The control steps for service restoration process of the example shown in Fig. 4.28

Service restoration process			Return process		
Set of the action nodes to control	Steps of control		Set of the action nodes to control	Steps of control	
	Control	Outage losses avoided ($)		Control	Outage losses avoided ($)
	Close13	108,109.64		Close7	131,818.36
	Close71	65,095.49		Close52	47,536.43
	Close72	47,063.29		Close54	43,345.54
	Open6	0		Open27	−4567.80
Action nodes to	Close4	88,421.37	Action nodes to open:	Close29	71,833.14
open:	Open49	0	6, 7, 18, 29, 52, 54,	Open51	−255.33
6, 49, 53, 66, 76	Close47	67,265.34	73, 74, 49, 53, 66, 76	Close53	42,089.36
	Open66	0		Open13	−108,109.64
Action nodes to	Close61	41,834.03	Action nodes to close:	Close18	108,109.64
close:	Open53	0	4, 13, 25, 27, 47, 51,	Open4	−88,421.37
4, 13, 25, 27, 47,	Close51	255.33	61, 71, 72	Close6	88,421.37
51, 61, 71, 72	Open76	−9086.76		Open71	−65,095.49
	Close25	9086.76		Close73	65,095.49
	Close27	4567.80		Open72	−47,063.29
				Close74	47,063.29
				Open25	−9086.76
				Close76	9086.76
				Open61	−41,834.03
				Close66	41,834.03
				Open47	−67,265.34
				Close49	67,265.34

The process from the fault isolation to the service restoration operation mode is called the *service restoration process*. The process from the service restoration to the normal operation mode after repair is called the *return process*.

The control steps of both service restoration and return process are more complicated with controls not only opening some switches but also closing some switches. The generation methodology of the control steps is the same for both processes, which will be described in the next subsection.

4.4.2.4.2 Generation of the Control Steps for Service Restoration Process and Return Process

The following principles should be followed in the control of both service restoration and return process:

- No loop is closed with damage. The transient closed loop is permitted in some cases, such as when the corresponding two feeders are from the same bus. It is not allowed in cases where there is risk of damage, such as too big a difference of voltage amplitude or when the phase between nodes opening the loop is much large.

- No overload is caused.
- Restore the loads with large outage losses as high priority. The outage losses may be measured by financial methods, which are sometimes rather difficult to obtain. For the sake of simplicity, load shedding (kW) may be used to represent outage losses instead.

The steps to generate the control steps for the service restoration and return process are as follows:

Step 1: Put the action nodes to be closed and opened into the queues of $\boldsymbol{Q}_c$ and $\boldsymbol{Q}_o$, respectively.

Step 2: Get an action node N_c to be closed from $\boldsymbol{Q}_c$. Search the upstream and the downstream of N_c. Once a source node, an ending node or an action node to be opened with tele-control (i.e., in Q_0) is met, the search of the corresponding direction is ended. If an action node to be open, N_o, is found, get N_o from $\boldsymbol{Q}_o$ and N_o and N_c to form a *class-2 operation group*, which is put into the queue of $\boldsymbol{G}_2$. If no action node to be opened is found, N_c forms a *class-1 operation group*, which is put into the queue of $\boldsymbol{G}_1$.

Step 3: If $\boldsymbol{Q}_c$ is not empty, return to Step 2.

Step 4: Calculate the outage loss to avoid for each class-1 operation group in $\boldsymbol{G}_1$, respectively. Arrange the control steps to guarantee the class-1 operation group avoids a large outage loss as high priority, and put the control steps according to priority into the queue ***SW***.

Step 5: Return to Step 4 until $\boldsymbol{G}_1$ is empty.

Step 6: Calculate the outage loss to avoid for each class-2 operation group in $\boldsymbol{G}_2$, respectively. Arrange the control steps to guarantee the class-2 operation group avoids a large outage loss as high priority, and put the control steps according to priority into the queue ***SW***. If the closed loop of a class-2 operation group is permitted, the close control is arranged before the open control, otherwise, the open control is arranged before the close control.

Step 7: Return to Step 6 until $\boldsymbol{G}_2$ is empty.

It is worth mentioning that a class-1 operation group contains only the controls to close action nodes while a class-2 operation group contains types of close control and open control. The class-1 operation groups are arranged in ahead of the class-2 operation groups.

An example of service restoration to avoid large area breakdown is shown in Fig. 4.28. The numbers beside the nodes are the corresponding sequence numbers. The data in brackets "( )" indicate the loads (A) distributed from the action regions. The rated current of the transformers, that is, the rated current of node-78, node-80, node-83, node-85, node-88, node-90, node-92, node-93, node-94, node-95, node-96, and node-97 are 1400 A. The rated current of the HV transmission lines, that is, the rated current of node-98, node-99, node-100, node-101, node-102, and node-103 are

1400 A. The #1 #5 HV transmission lines are double circuit lines implanted on the same tower. The #3 and #6 HV transmission lines are double circuit lines implanted on the same tower. The #2 and #4 HV transmission lines are double circuit lines implanted on the same tower. The rated current of the other action nodes is 400 A. Figure 4.28 shows the normal operation mode with total distributed load of 3072 A. The outage losses of the action regions are listed in Table 4.14. Supposing that a tower of #2 HV and #4 HV transmission lines collapses causing both Bus-5 and Bus-6 to lose voltage, the remote control steps of the service restoration and return process are shown in Table 4.15. The total outage losses costing $422,612.28 can be avoided.

4.4.2.5 Correction Control in Cases of Remote Control Failure in Service Restoration Processes

If the remote control of action nodes is successful, the service restoration program ends. Otherwise, we should carry out the re-control process to control the failed action nodes three times. If all the re-control processes are successful, the service restoration program ends. Otherwise, we should carry out the correction control process of service restoration as follows:

Step 1: If all of the restorable regions are restored, the correction control process of service restoration ends; otherwise, go to Step 2.
Step 2: Mark the action nodes that failed to control and let them keep their current state without taking part in the service restoration correction control process.
Step 3: Mark the action nodes that successfully controlled and let them keep the current state without taking part in the service restoration correction control process.
Step 4: Find out the new optimal service restoration scheme, in which the solution space consists of the action nodes without marks.
Step 5: Generate new steps of action nodes to control in the service restoration correction control process.

An example is shown in Figure 4.29. Suppose that no closed loop is permitted. A permanent fault occurs in Re(A_2,A_3).

S_1 is tripped by the relay protection device immediately. Then action nodes A_2 and A_3 are opened by remote control to isolate the fault, which is shown in Figure 4.29(b).

The optimal service restoration steps are as follows:

- Close the action node A_6 to restore the loads downstream of the fault region.
- Close the action node S_1 to restore the loads upstream of the fault region.
- When the optimal service restoration is completed, the LBD should be 0.28.
- Assume that the action node-A_6 fails to control to the closed state. Thus, we should carry out the corrected control process of service restoration. The corrected control steps are as following:

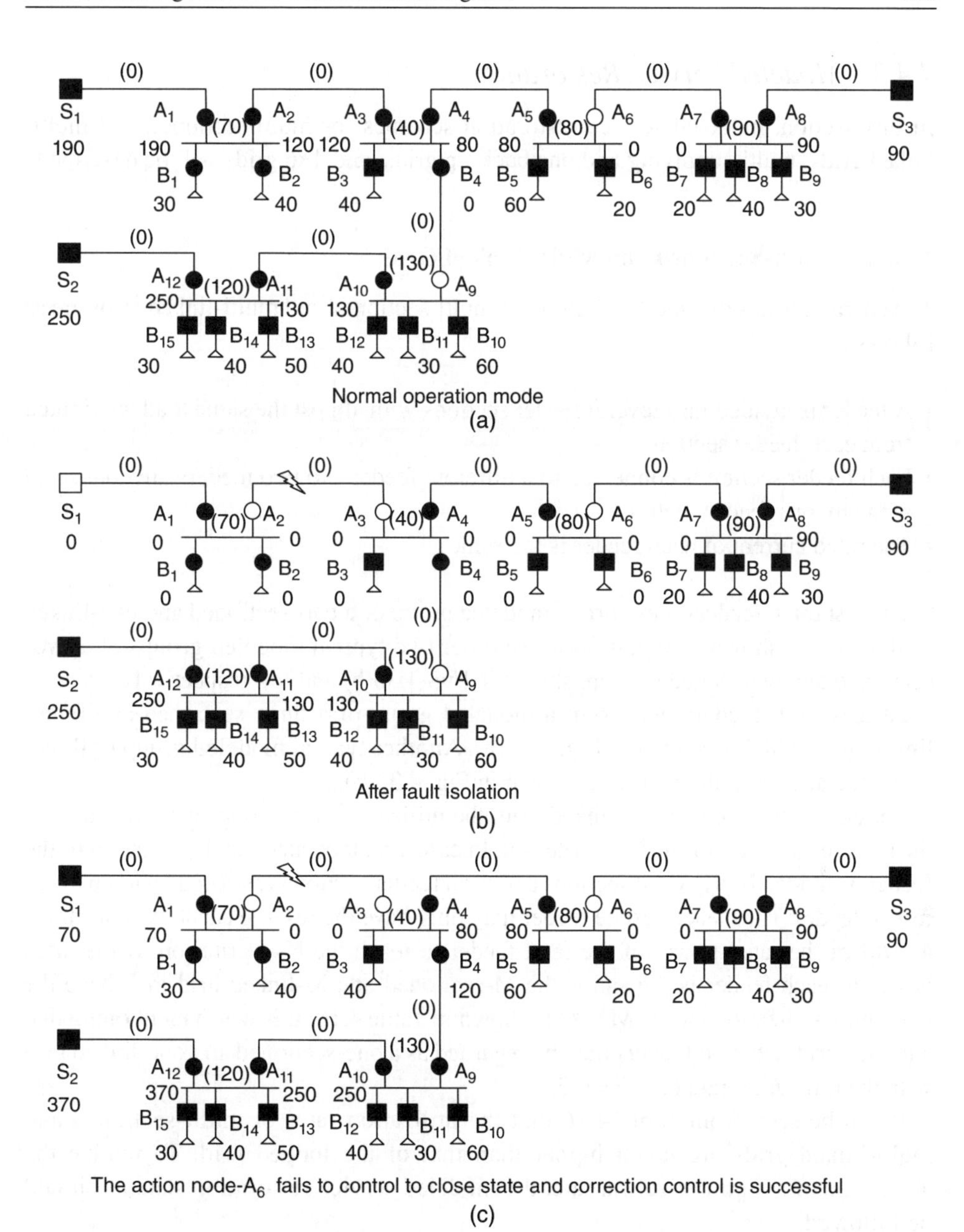

Normal operation mode
(a)

After fault isolation
(b)

The action node-A_6 fails to control to close state and correction control is successful
(c)

Figure 4.29 An example of correction control in a service restoration process

Close the action node A_9 to restore the loads downstream of the fault region.
Close the action node S_1 to restore the loads upstream of the fault region.

Suppose that the correction controls are all successful as shown in Figure 4.29(c). Although there is no load shedding due to overload, the LBD decreases to 0.18.

4.4.3 Modeled Service Restoration

In this section, modeled service restoration schemes for multi-sectioned and multi-linked grids, multi-supplying and one back-up grids, and 4×6 grids will be described.

4.4.3.1 Multi-Sectioned and Multi-Linked Grid

The characteristics of a modeled group of multi-sectioned and multi-linked grids are as follows:

- A feeder is divided into several feeder sections with almost the same load distributed from each feeder section.
- Each feeder section is connected to a different feeder. No two feeders are connected with any one feeder section.
- The rated current of each feeder is the same.

At least three feeders may form a modeled group of a two-sectioned and two-linked grid, which is shown in Fig. 4.30(a). Another two typical modeled groups of a two-sectioned and two-linked grid are shown in Fig. 4.30(b) and (c), respectively.

At least four feeders may form a modeled group of a three sectioned and three-linked grid, which is shown in Fig. 4.32(a). Another typical modeled group of three-sectioned and three-linked grid is shown in Fig. 4.31(b).

For an M-sectioned and M-linked grid, the utilization rate is improved by reducing the reserve capacity for the N−1 criterion. In case a fault occurs, the downstream of the feeder is divided into several sections and each feeder section is restored by a separated rescue feeder. Thus, each rescue feeder may only have the reserve capacity of as much as 1/M of the rated power of the fault feeder to reach the N−1 criterion. If the rated power of each feeder is the same, the M-sectioned and M-linked grid may have the maximum utilization rate of M/(M+1) shown in Table 4.16. It is worth mentioning that a looped grid with two feeders may be regarded as a one-sectioned and one-linked grid with the utilization rate of 1/2 = 0.5.

It can be seen from Table 4.16 that the utilization rates of multi-sectioned and multi-linked grids are much higher than that of the looped grid. To realize the advantage of a high utilization rate, the modeled service restoration scheme should be followed.

The modeled service restoration scheme of a three-sectioned and three-linked grid is used as an example shown in Fig. 4.32. The feeder with the main source node S_1 is divided into three feeder sections by the action nodes of A and B. The three feeder sections are connected to the feeders with the main source nodes of S_2, S_3, and S_4 by the action nodes C, D, and E, respectively, which is shown in Fig. 4.32(a).

If a fault occurs in Re(B,C), Re(S_1,A, D) and Re(A,B,E) are restored by S_1, as shown in Fig. 4.32(b).

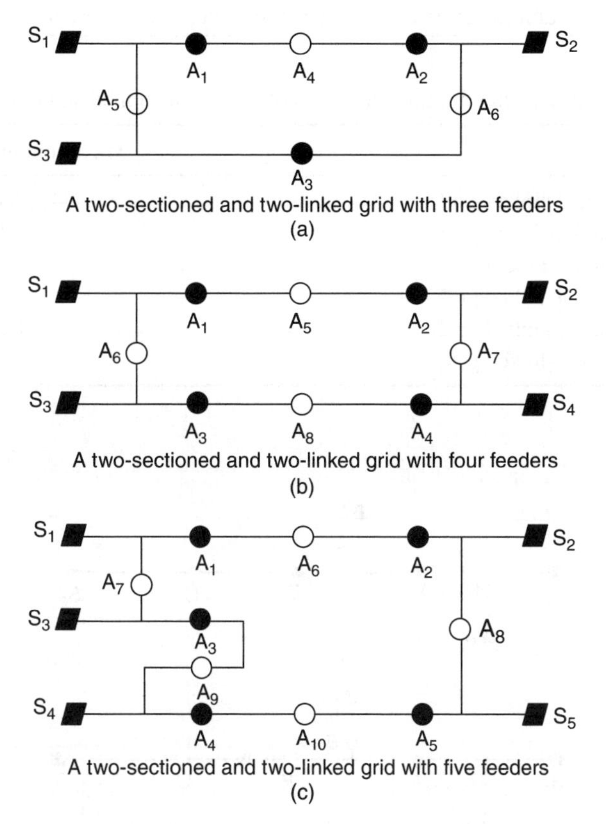

Figure 4.30 Typical two-sectioned and two-linked grids

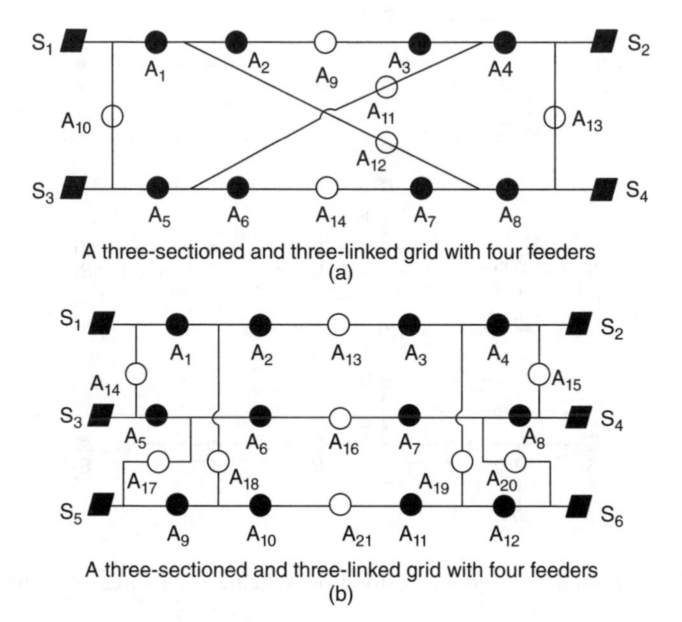

Figure 4.31 Typical three-sectioned and three-linked grids

Table 4.16 The maximum utilization rates of a multi-sectioned and multi-linked grid

Types	Maximum utilization rate
1-sectioned and 1-linked grid	50%
2-sectioned and 2-linked grid	67%
3-sectioned and 3-linked grid	75%
4-sectioned and 4-linked grid	80%
5-sectioned and 5-linked grid	83%
6-sectioned and 6-linked grid	86%

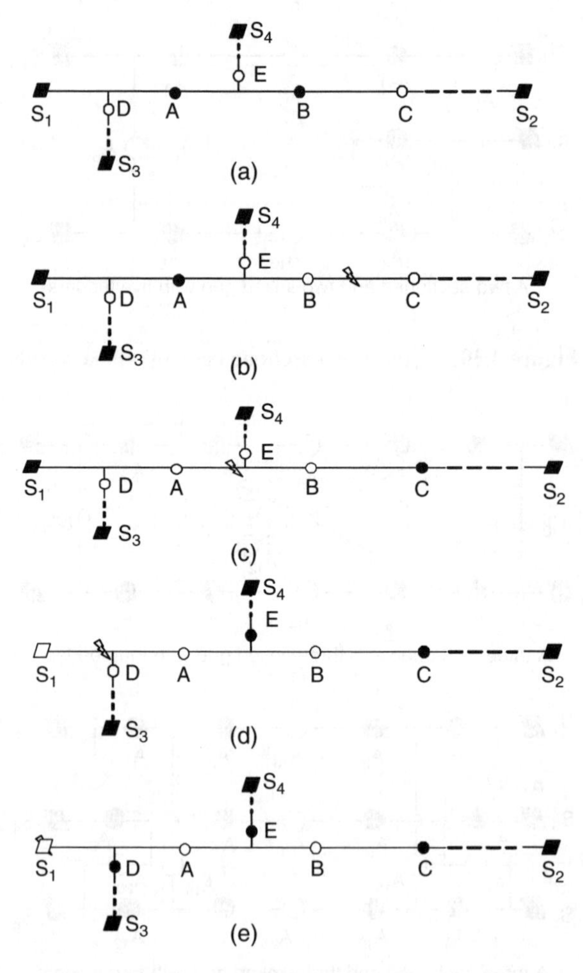

Figure 4.32 The modeled service restoration scheme of a three-sectioned and three-linked grid

If a fault occurs in Re(A,B,E), Re(S_1,A, D), is restored by S_1 and Re(B,C) is restored by S_2, respectively, as shown in Fig. 4.32(c).

If a fault occurs in Re(S_1,A, D), Re(A,B,E) is restored by S_4 and Re(B,C) is restored by S_2, respectively, as shown in Fig. 4.32(d).

In case of S_1 failure, Re(S_1,A, D) is restored by S_1, Re(A,B,E) is restored by S_4, and Re(B,C) is restored by S_2, respectively, as shown in Fig. 4.32(e).

4.4.3.1.1 Multi-Supplying and One Back-Up Grids

The characteristics of a multi-supplying and one back-up grid are as follows:

- A group of feeders are connected to one bus.
- In the normal operation mode, one of the feeders is idle while the other feeders may carry loads no greater than their rated current.
- The rated current of each feeder is the same.

A two-supplying and one back-up grid and a three-supplying and one back-up grid both based on cables are shown in Fig. 4.33(a) and (b), respectively.

Since one of the feeders is idle while the other feeders may carry the loads up to their rated current in normal operation mode, an M-supplying and one back-up grid may have the maximum utilization rate of M/(M+1) shown in Table 4.17.

It is worth mentioning that the grid shown in Fig. 4.34 does not meet the requirement of two-supplying and one back-up grid because no idle feeder exists. Supposing that the rated currents of the three feeders are the same at 600 A. The loading current of the feeder with the main source node of S_3 is 200 A. Thus, the maximum loading current of the other two feeders may be no more than 400 A. Otherwise, the N-1 criterion cannot be satisfied. Therefore, the maximum utilization rate is as low as 56%, which is much lower than that of the two-supplying and one back-up grid.

The modeled service restoration scheme of two-supplying and one back-up grid is used as an example shown in Fig. 4.35.

Whenever any fault occurs on one of the three feeders with loading shown in Fig. 4.35(a), the back-up feeder will provide the power to the downstream of the fault region, while the upstream will be restored by the main power source.

In the case of the fault being in Re(A_4,A_5) shown in Fig. 4.35(a), S_1 is tripped to clear the fault current, as shown in Fig. 4.35(a).

Action nodes A_4 and A_5 are open by remote control to isolate the fault region, as shown in Fig. 4.35(b).

S_1 is closed by remote control to restore the upstream of the fault region, as shown in Fig. 4.35(c).

Action node A_8 is closed by remote control to restore the downstream of the fault region, as shown in Fig. 4.35(d).

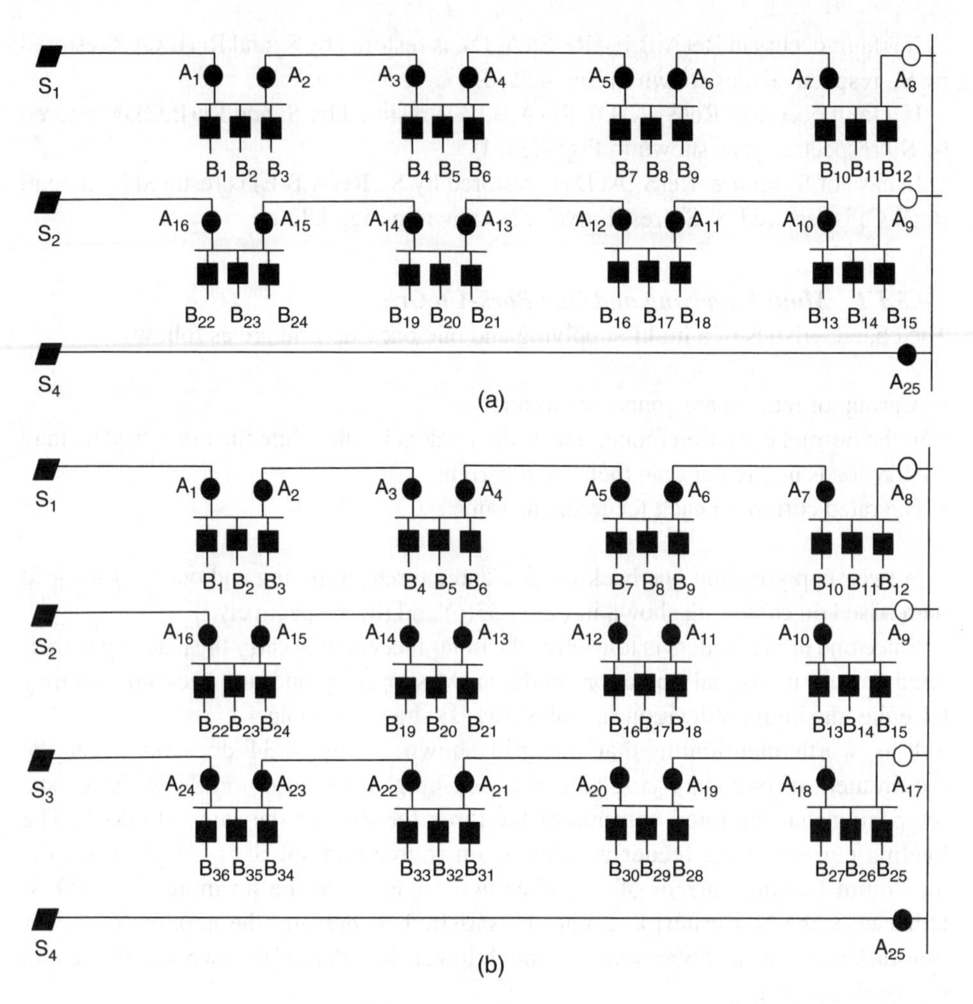

Figure 4.33 Typical multi-supplying and one back-up grids

Table 4.17 The maximum utilization rate of M-supplying and one back-up grids

Types	Maximum utilization rate
two-supplying and one back-up	67%
three-supplying and one back-up	75%
four-supplying and one back-up	80%
five-supplying and one back-up	83%
six-supplying and one back-up	86%

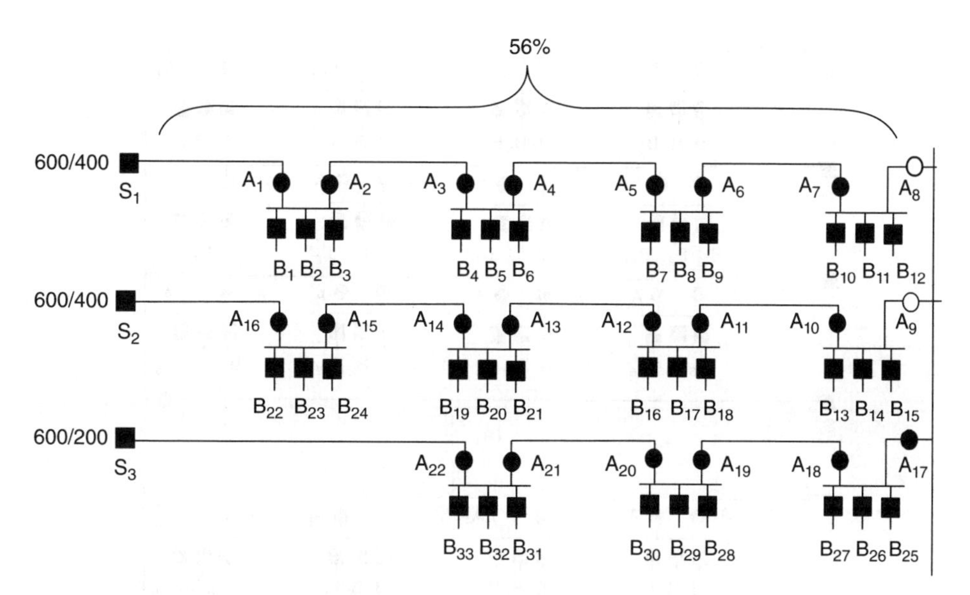

Figure 4.34 Without the idle feeder, the grid is not a two-supplying and one back-up grid

Although sometimes the loading is so light that the downstream of the fault region Re(A_4,A_5) can be restored by main power source S_2, the restoration scheme is always to restore the downstream of the fault region by the back-up feeder.

4.4.3.1.2 The 4 × 6 Connection Grid

A 4 × 6 connection grid is shown in Fig. 4.36(a), in which the transformers are expressed by their secondary switches S_1, S_2, S_3, and S_4, respectively. The characteristics of a 4 × 6 connection grid are as follows:

- Four transformers and six feeders form a modeled group of a 4 × 6 connection grid, in which each pair of transformers is linked by an open looped feeder pair containing two feeders connected with a loop switch.
- Each open looped feeder pair meets the requirement of the N−1 criterion.
- The rated power of each transformer is the same.
- The rated current of each feeder is the same.

Since the six feeders form three open looped feeder pairs, each feeder has a maximum utilization rate of 50%, which is no higher than that of the looped grid. But each transformer may only have the reserve capacity of one third its rated power because once a transformer fails the other three transformers may share its loading in the restoration process. Thus, each transformer has a maximum utilization rate of 75%.

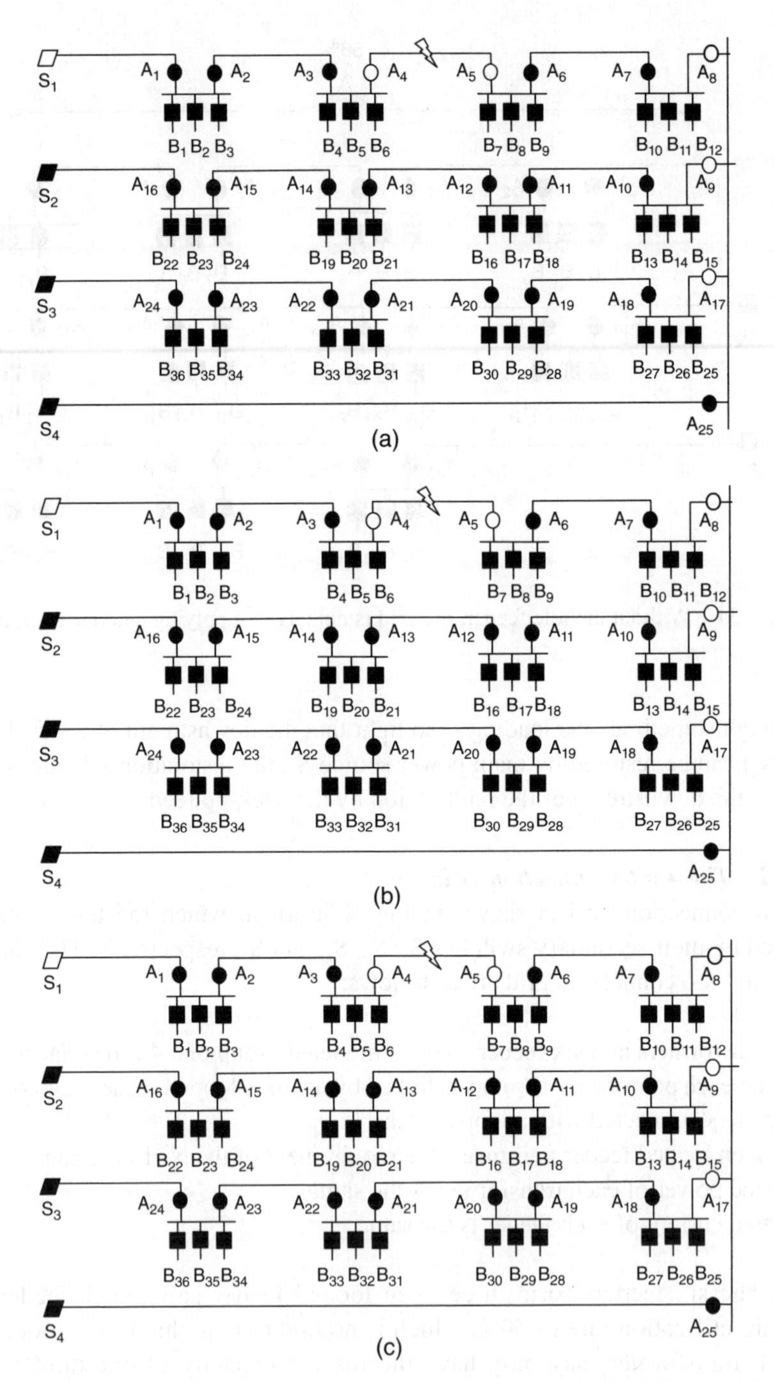

Figure 4.35 The modeled service restoration scheme of three-supplying and one back-up grid

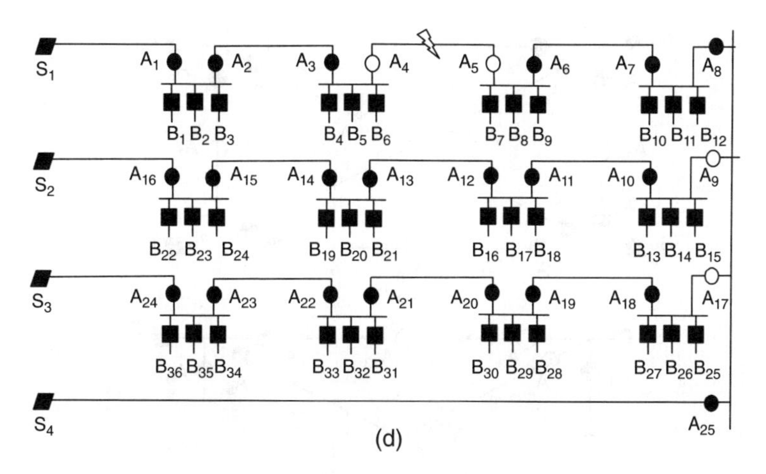

Figure 4.35 *(continued)*

The modeled restoration principles of the 4 × 6 connection grid shown in Fig. 4.36(a) are:

- When a fault occurs in the downstream region of a feeder, such as the fault being in Re(A_3,A_{13}), the fault current is cleared by tripping S_1 and the fault is isolated by opening the action node A_3 by remote control. Then the upstream region is restored by closing the main source node S_1 by remote control, which is shown in Fig. 4.36(b).
- When a fault occurs in the upstream region of a feeder, such as the fault being in Re(S_1,A_1,A_2,A_3), the fault current is cleared by tripping S_1 and the fault is isolated by opening the action nodes A_1, A_2, and A_3 by remote control. Then the downstream regions are restored by closing the action nodes A_{13}, A_{14}, and A_{15} by remote control, which is shown in Fig. 4.36(c).
- When a fault occurs on a transformer, such as the transformer being connected to S_1, the fault current is cleared by tripping S_1 and the fault is isolated by opening the action nodes A_1, A_2, and A_3 by remote control. Then the downstream regions are restored by closing the action nodes A_{13}, A_{14}, and A_{15} by remote control, which is shown in Fig. 4.36(d).

4.4.4 Coordination of the Four Types of Service Restoration

The coordination of short time service restoration, real time service restoration, service restoration with minimum load shedding, and modeled service restoration is shown in Fig. 4.37.

It can be seen from Fig. 4.37 that modeled service restoration applies to the corresponding modeled group. If some action nodes fail to control in a modeled service

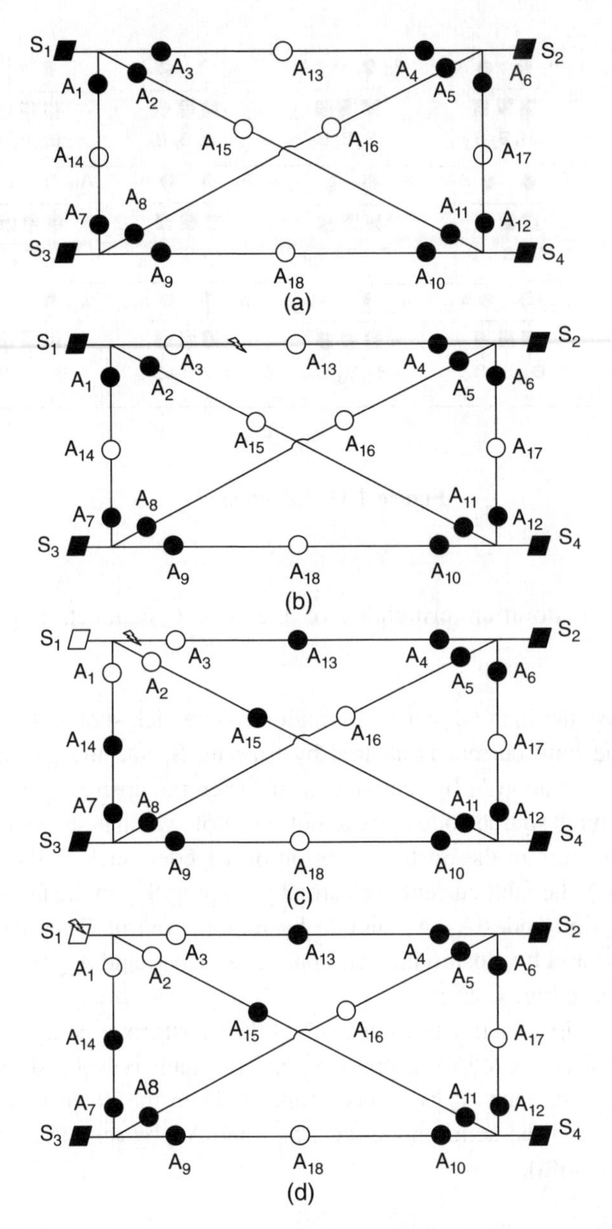

Figure 4.36 The modeled service restoration scheme of a 4 × 6 connection grid

restoration, short-time, real-time, or service restoration with minimum load shedding may take place, among which, short-time service restoration is of the highest priority, followed by real-time. Service restoration with minimum load shedding is the last choice.

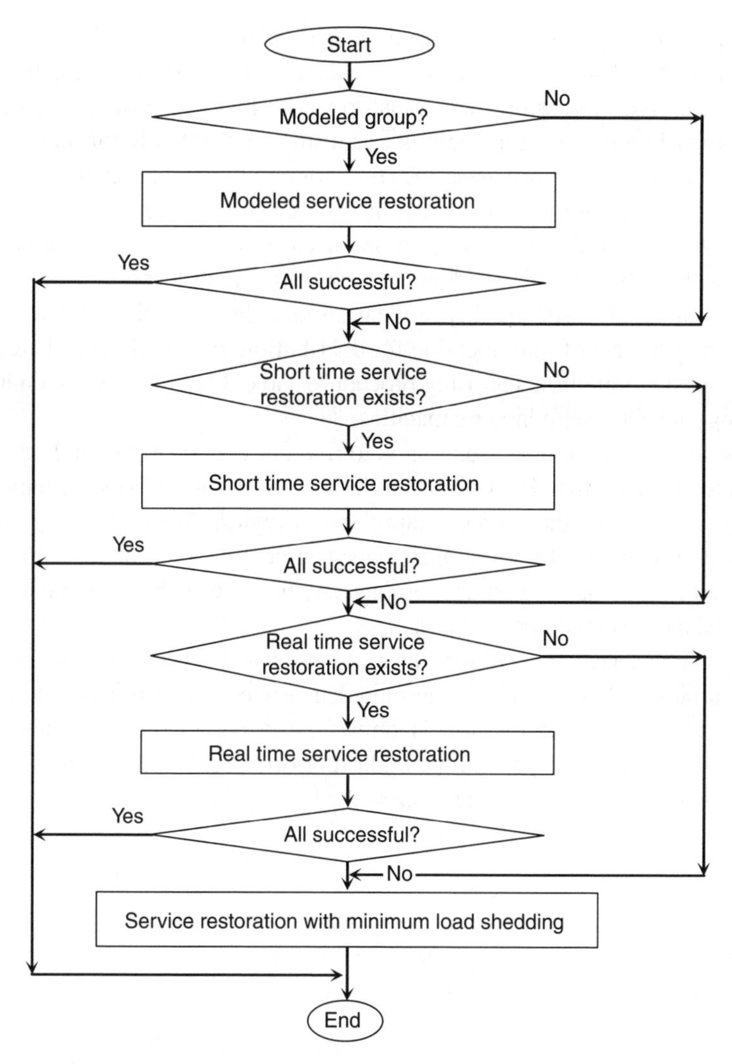

Figure 4.37 Coordination of the four types of service restoration

4.5 Conclusion and Summary

The fault processing program of a DAS is composed of the fault information collecting process, the fault location process, and the fault isolation and service reiteration process.

The simplified model of distribution grid is effective for fault processing. Nodes can be classified into four types; action, monitoring, ending, and source nodes. An action region is the smallest unit of fault isolation and load transfer. A monitoring region is the smallest unit of fault location. A connected system is the smallest unit for network reconfiguration and service restoration. A modeled group is the sub-graph containing a certain intentionally designed connection. A special type of modeled group is the conscious island with some DGs and loads inside. Nodes and regions can be described in hierarchies. Node-to-region load transform describes the program of calculating the loads distributed from the regions according to the loads through the nodes. Region-to-node load transform describes the program of calculating the loads flowing through the nodes according to the loads distributed from the regions.

Criteria 4.1–4.5 apply to fault location in cases of sufficient information. The Bayesian analysis based fault location approach is for fault location in cases of insufficient information. The adaptability of the traditional fault location approaches is high in solving the problem of fault location for distribution grids with DGs. Fault location based on coordination of the reclosing procedure and cut out of DGs is a fault location methodology for DGs with larger capacities.

The first thing to do in fault isolation is to find out the corresponding action region of the located fault region. Then carry out the remote control process of fault isolation to trip the terminals of the corresponding action region. For distribution grids with DGs, service restoration by the main power sources is of the highest priority, after which some of the remaining outage regions may be restored by DGs according to the conscious islands arranged beforehand.

The short-time service restoration, the real-time service restoration, service restoration with minimum load shedding and modeled service restoration should be coordinated. Modeled service restoration is only for the corresponding modeled group. Short-time service restoration is of the highest priority, followed by real-time. Service restoration with minimum load shedding is the last choice.

5

Single Phase to Ground Fault Processing

Dong Xinzhou and Shi Shenxing

Abstract

Chapter 5 introduces different single-phase-to-ground fault characteristics in power distribution systems with different types of neutral grounded, focuses on high impendence fault in power distribution systems with neutral effectively grounded and single-phase-to-ground fault in power distribution system with neutral non-effectively grounded. It proposes high impendence fault detection methods, presents single-phase-to-ground fault feeder selection, and protection, prevention, and location of single-phase-to-ground faults.

Keywords

single-phase-to-ground fault, power distribution system, neutral grounded non-effectively, high impedance(resistance) fault detection, third harmonic current, fault feeder selection, fault location, traveling waves, zero-sequence current, fault prevention

A ground fault means there is an abnormal connection between wire and ground, including single- and two-phase to ground faults. According to statistics, the number of single-phase-to-ground faults accounts for more than 90% of the total number of faults in transmission lines and more than 80% of the total number of faults in distribution lines. Most phase-to-phase faults are evolved from single-phase-to-ground fault. This chapter mainly discusses the single-phase-to-ground fault and its protection technologies.

Fault Location and Service Restoration for Electrical Distribution Systems, First Edition. Jian Liu, Xinzhou Dong, Xingying Chen, Xiangqian Tong, Xiaoqing Zhang and Shiming Xu.

5.1 Types of Ground Fault and Protection Strategy

The ground fault is closely related to the method of neutral grounding. For the same fault conditions but different methods of neutral grounding, a ground fault shows different characteristics, effects, and harm. Accordingly, the protection strategies are different.

5.1.1 The Neutral Grounding Mode and Ground Fault Types

There are several classification methods to the neutral grounding modes:

- Grounded and non-grounded, according to whether the neutral point is connected to the ground.
- According to the magnitude of single-phase-to-ground fault current, they are divided into neutral effectively and neutral non-effectively grounded, as well as high- and low-current grounding.
- According to the situation of the connection, they are divided into neutral directly and neutral indirectly grounded.

From the system point of view, according to the different neutral grounding modes, the systems are called neutral grounded, neutral non-grounded, neutral effectively grounded, neutral non-effectively grounded, neutral directly grounded, and neutral indirectly grounded.

The selection of neutral grounding modes depends on two factors: power supply reliability and limitation of overvoltage.

Figure 5.1 shows neutral grounding modes of three different classification methods. Figure 5.2 shows a directly grounded system, ground point, ground and neutral point, and phase conductor that make up a short circuit path, so there is high current passing through the fault phase. In order to protect the failure equipment, the breaker must take action to cut off the failure path. Considering the probability of a single-phase-to-ground fault, power supply reliability is lowest. On the other hand, when the single-phase-to-ground fault occurs in this system, its grounded phase voltage decreases, and sound phase voltages are almost the same, while grounded phase current increases and sound phase currents are almost the same. So overvoltage problems can be ignored in this mode, but the fault must be cleared.

In Fig. 5.3 showing the low-resistance grounding system, the resistance between the neutral point and ground limits the ground fault current and overvoltage after the fault occurs. It is a mode that is commonly used abroad and begins to be used in China, which belongs to the neutral effectively grounded system. There is still high ground fault current after a fault occurs, so the breaker must immediately take action to cut off the grounded path, which also leads to the interruption of power supply.

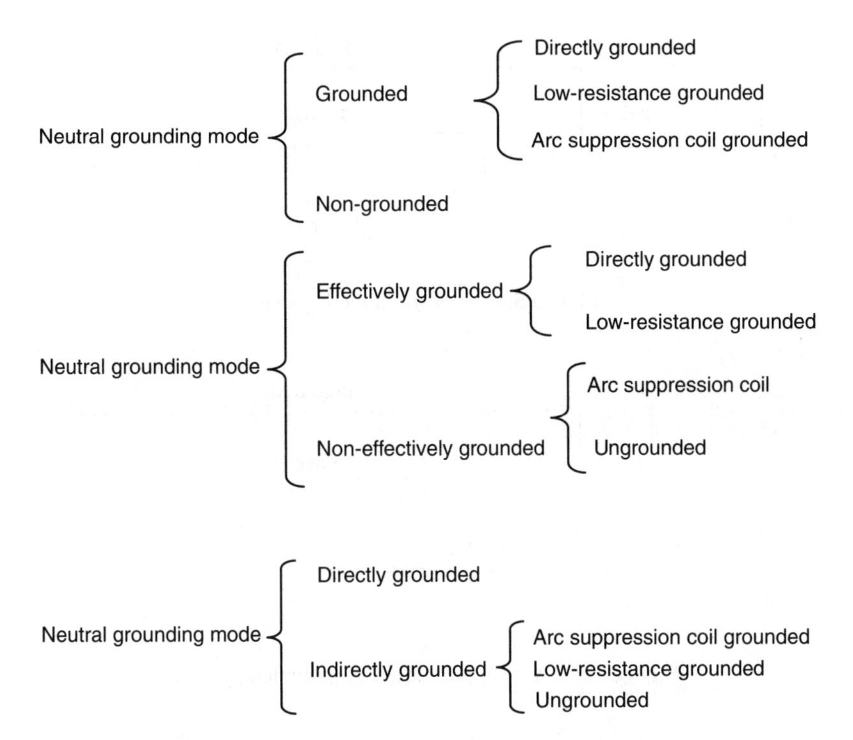

Figure 5.1 Classification of neutral grounding mode

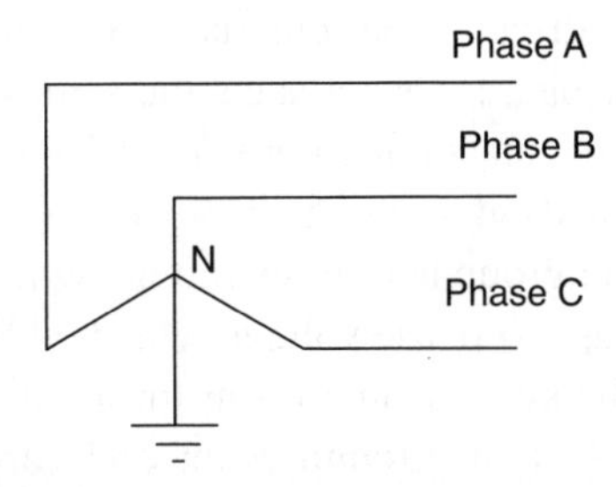

Figure 5.2 Neutral directly grounded

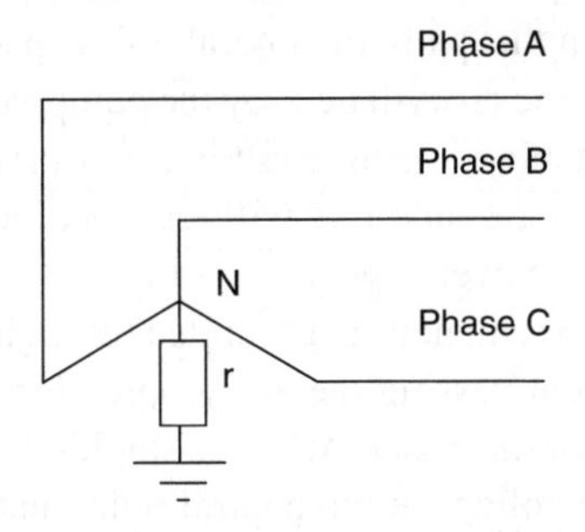

Figure 5.3 Neutral grounded through low-resistance

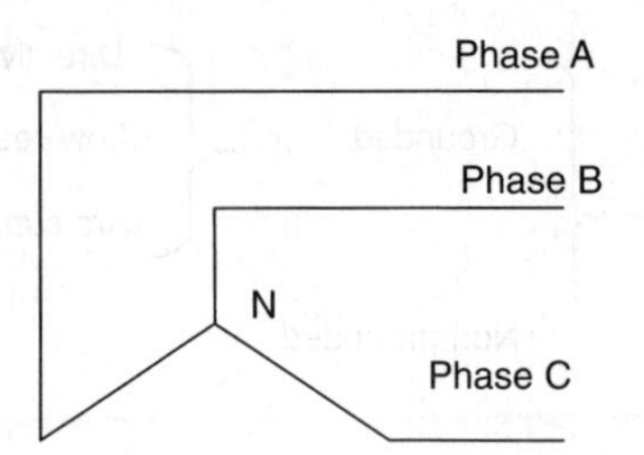

Figure 5.4 Non-grounded system

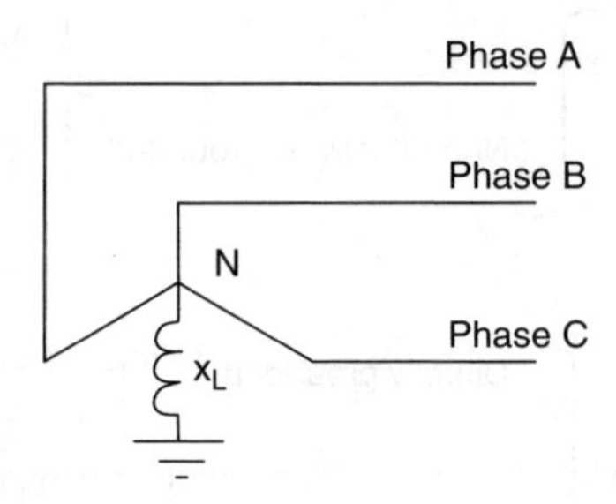

Figure 5.5 Arc suppression coil grounded system

As for the non-grounded system showed in Fig. 5.4, when a single-phase-to-ground fault occurs, a short circuit path cannot be constructed for a neutral point non-grounded. The fault phase and healthy phase current stays the same and the line voltage maintains symmetry, so the system can maintain operation for a short time. We can use this time to figure out the reason and cut off the fault, or we can switch load, so the reliability of this mode is high. But the grounded phase voltage will decrease and sound phase voltages will increase to phase-to-phase voltage, which will threaten the insulation of electrical equipment, and the system cannot run for a long time after a fault occurs. In fact, for non-grounded system, ground point and capacitance between ground and conductor can make up a circuit because of the existence of line capacitance (the capacitance is small, but capacitive reactance is high), and in this way, there is low capacitive current passing in this path. In general, this capacitive current acts as an arc and the high temperature of the arc will destroy the equipment, setting fire to the architecture nearby. And an unstable arc will result in arc overvoltage, causing breakdown of the insulation of sound phase, and this will lead to a phase-to-phase fault making the breaker act and interrupt power supply.

Figure 5.5 shows a system with neutral grounded through an arc suppression coil. In this figure, there is no current passing the arc suppression coil between neutral point and ground, and the coil doesn't work. When a single-phase-to-ground fault occurs, there will be zero sequence voltage at the neutral point and there will be reactive current passing through the coil into the grounded system to counteract the capacitive ground current passing the ground point, reducing the harm caused by the ground arc current. To be sure, there will be no capacitive current or low current passing the neutral point when using the arc suppression coil. But the system is indeed grounded, so the ground fault may still occur. And the grounded phase voltage will decrease but sound phase voltage will still be very high. So the system is not allowed to run for a long time.

The fault characteristics here are distinguished by the structure of the neutral point. In fact, the size and characteristics of ground current are also influenced by the fault point. This point may be metallic or nonmetallic grounded. In general, a nonmetallic grounded situation includes grounding through an arc, tree branch, tower, and all or part of them. A fault via a nonmetallic medium is often called a high resistance fault. Its characteristic is low fault current, which is hard to measure.

5.1.2 The Protection Strategies for Different Types of Ground Faults

Based on these statements, when a single-phase-to-ground fault occurs in a neutral effectively grounded system, we need to find out and cut off the failure path immediately, so appropriate relay protection is necessary. Fortunately, zero sequence voltage and current will appear after a ground fault occurs, which is a very significant characteristic. Based on this, we can develop single-phase-to-ground protection based on zero sequence voltage and current. It is even more sensitive than over-current protection and directional over-current protection used to protect phase-to-phase fault, for they need to be distinguished in a heavy load situation, while zero sequence voltage and current based protection don't need this. But when high resistance ground fault occurs in a low resistance grounded system, the fault current is low, and it's hard to protect such a fault because of the high resistance.

Unlike the effectively grounded system, when single-phase-to-ground fault occurs in the non-effectively grounded system, the fault current is commonly almost zero and the fault characteristics are not obvious, except for the appearance of zero sequence voltage, for example, a single-phase-to-ground fault in a short feeder (the distributed capacitance and current is small) of the non-grounded system, or a single-phase-to-ground fault in the system with neutral grounded through an arc suppression coil. Fortunately, this situation doesn't affect the power supply and causes little direct harm to the system. But operators have to know that (1) a ground fault occurs and (2) where a fault occurs, after it occurs. That is to say, protection for single-phase-to-ground fault is necessary. And the purpose is to alarm but not trip. This has promoted another technology: single-phase-to-ground fault line selection technology in neutral non-effectively grounded systems. Neutral non-effectively grounded is widely used in power distribution networks in China. And in some developed coastal cities, low resistance grounded systems are acceptable for avoiding operating with faults after a single-phase-to-ground fault occurs in the system. This chapter mainly discusses the measurement and technology for faults in neutral non-effectively grounded systems. It also introduces the technology and method of fault detection of high resistance faults in power distribution systems with neutral grounded and through low resistance.

5.2 Detection of High Resistance Ground Faults in Low Resistance Grounded Systems

5.2.1 High Resistance Ground Faults

High resistance ground faults are faults where a conductor is connected with ground via a nonmetallic conducting medium. These media include roads, soil, tree branches, and concrete buildings. Its main characteristic is that a nonmetallic conducting medium has high resistance, leading to a low ground fault current. The fault has the characteristics of arcing, and being intermittent and instantaneous, which is hard to detect for traditional zero sequence current protection.

High resistance ground faults occur in overhead lines and cable lines under all kinds of voltage levels. When a single-phase-to-ground fault occurs in 220–500 kV power lines, high resistance often exists. When conductors discharge to tree branches, fault resistance may rise to 100–300 Ω. Also, such faults also occur in 6–35 kV power distribution systems with neutral ungrounded, grounded through arc suppression coils, or grounded through low resistance.

When a single-phase-to-ground fault occurs in the system with neutral grounded through low resistance, the resistance limits the fault current. This resistance is between the neutral point and ground, and is in series in the circuit, making the fault current decrease.

The single-phase-to-ground fault in the power distribution system with neutral non-effectively grounded has similar characteristics to a high resistance fault in the system with neutral grounded directly, for both resistance at the neutral point and high resistance at the fault point make the fault current low. When both of them exist, the fault current will get smaller, so this chapter doesn't distinguish between them and discuss them together.

It is commonly accepted internationally (such as via the IEEE and PSERC, Power System Engineering and Research Center) that high resistance faults particularly refer to faults in the system with neutral effectively grounded (like four-wire system in north America) via a nonmetallic conducting medium, and the fault current is lower than threshold value of the over-current protection. Accordingly, the protection will not work for it.

To all kinds of high resistance fault, the fault current always is low. Table 5.1 shows high resistance fault currents in a 12.5 kV (12.2 kV to ground) system with neutral grounded directly, given by PSERC. In general, the high resistance fault current is lower than 50 A, lower than the lowest over-current protection operational value.

Although the current is small, the fault occurs. In the system with neutral grounded effectively, the fault must be cleared. In the system with neutral grounded

Table 5.1 Typical values of high resistance fault currents in grounded systems

Medium	Current(A)
Dry pitch/concrete/sand	0
Wet sand	15
Dry turf	20
Dry grassland	25
Wet turf	40
Wet grassland	50
Reinforced concrete	75

non-effectively, the alarm signal must be issued by protection. At present, the protections of high resistance faults are:

- Zero sequence inverse-time over current protection, a method to detect the zero sequence current by setting the operating value of protection as low. Reliability and selectivity are guaranteed by delaying the operation time.
- Protection based on the third harmonic current. Detecting relations of the amplitude and phase of the third harmonic current can make up protection, because nonlinear elements like transition resistance and a fault arc will lead to a third harmonic current.
- Protections using superimposed components of sampling value. This protection detects high frequency disturbance caused by high resistance fault and the gist is that the high resistance fault arc can produce high frequency noise.

5.2.2 *Zero Sequence Inverse-Time Overcurrent Protection*

In order to improve the sensitivity of the fault detection, the relay startup current is set to avoid the normal unbalanced current I_{ub}, it can be set as:

$$I_{K\cdot act} = \frac{K_{rel}}{K_{re}} I_{ub} \tag{5.1}$$

A long operation time delay should be chosen. $I_{K\cdot act}$ is the startup current of the relay, I_{Ub} is the normal unbalanced current, K_{rel} is the startup coefficient, K_{re} is the return coefficient.

The action characteristic of the relay often uses high inverse-time characteristics, and the equation is:

$$t = \frac{120K}{\left(\dfrac{I}{I_{K\cdot act}}\right) - 1} \tag{5.2}$$

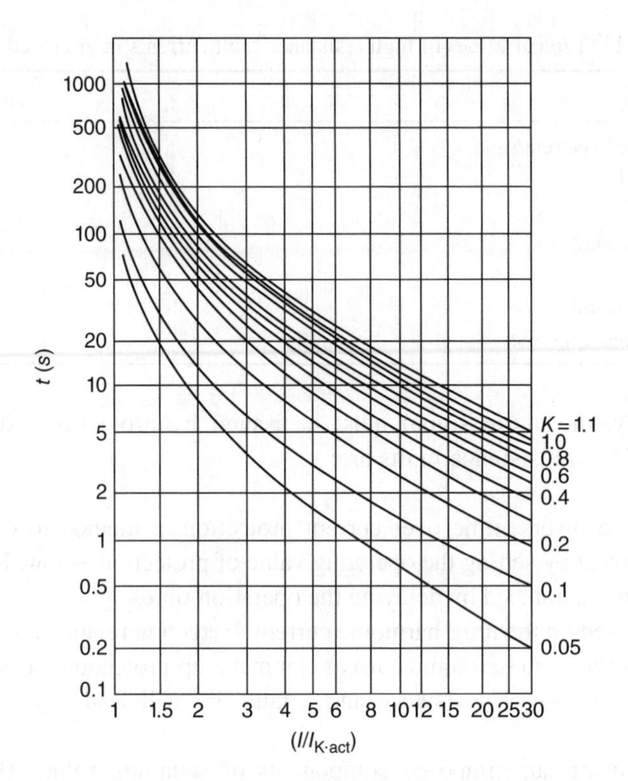

Figure 5.6 Current-time characteristic curve of high inverse-time current relay

Where, $I_{K \cdot act}$ is the startup current of relay, I is the current passing the relay, K is the time setting coefficient, t is the action time, and its characteristic curve is showed in Fig. 5.6.

Because the startup current is adjusted to a low value, when high unbalanced current appears under phase-to- phase fault outside the protected line and two-phase operation after the single phase circuit breaker opens, the relay will start. At this time, we have to adjust the long time delay to guarantee selectivity and avoid the malfunction.

5.2.3 Grounded Protection Based on the Amplitude and Phase of the Third Harmonic Current

Research in other countries besides China, especially on single-phase-to-ground fault protection in power distribution systems with neutral grounded through low resistance, mainly focuses on the detection of fault voltage and current waveform characteristics. Among these detection methods based on waveform characteristics, the third harmonic current based method is used.

The principle of this method is that when single-phase-to-ground fault through high resistance occurs in the distribution line, the fault current's waveform will warp because

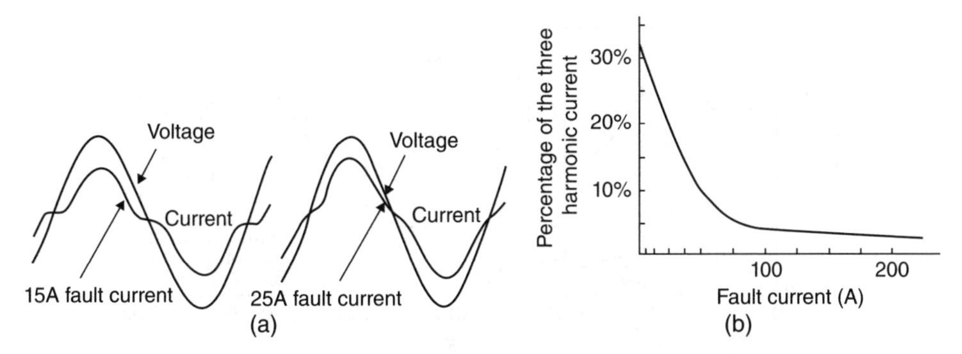

Figure 5.7 Waveform of high resistance fault current and (a) third harmonic current; (b) ratio of third harmonic current in the fault current

of the nonlinear grounded resistance, which comes from the arc and the nonlinear character of a compound itself in a medium such as soil (e.g., carborundum). This is because when a high resistance ground fault occurs, the fundamental current is low but the nonlinear character is obvious, which will generate third harmonic signals. Meanwhile, considering the transformers connected at angles in power distribution systems will cut off a zero sequence third harmonic circuit, which will make fault distribution lines become relatively independent paths for the third harmonic current. So, the third harmonic is called a typical harmonic characteristic of a high resistance fault.

Figure 5.7(a) shows the typical waveform of the fault voltage and current of single-phase-to-ground fault. Figure 5.7(b) shows the third harmonic current in the fault current. Many field measurements get similar results.

Analyzing the waveform in Fig. 5.7(a), we can find that, at the fault point, the phase of third harmonic current is the function of instantaneous value of system voltage. Because of the nonlinear characteristics of resistance, the fault current appears in the shape of a "peaked wave". So we can assume that at the fault point, when fundamental voltage rises to peak value, the third harmonic current and fundamental current rises to the maximum value too. Then, at the zero crossing point of the third harmonic current, the phase-angle difference between third harmonic voltage and fundamental voltage is 180° (the rotational speed of the third harmonic is three times fundamental voltage). Figure 5.8 shows the waveform decomposition.

It should be noted that the measurement of third harmonic current is based on its increment for existence of background harmonics and current analysis is carried out using a superimposed circuit.

From the figure, we can see that at the zero crossing point of the third harmonic current, the third harmonic current and the voltage have phase angle difference of 180°, which results from the nonlinear characteristics of fault arc and resistance at the fault point. These are also regarded as the characteristics of a high resistance ground fault. But the phase relations will be changed as measured at the bus bar in the power substation, as shown in Fig. 5.9. This change can be determined with the known system

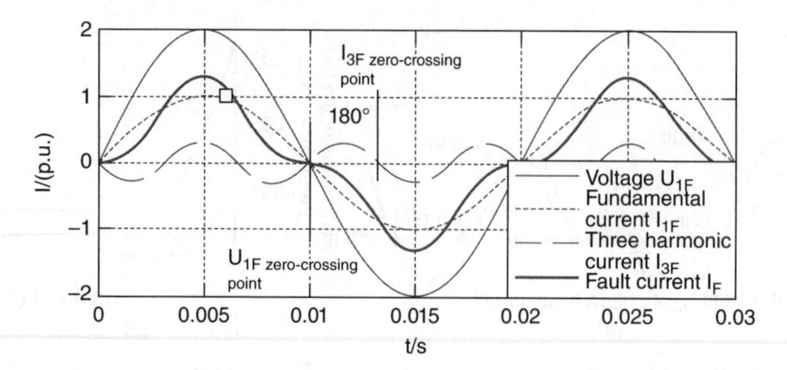

Figure 5.8 Voltage, current, and current decomposition waveforms

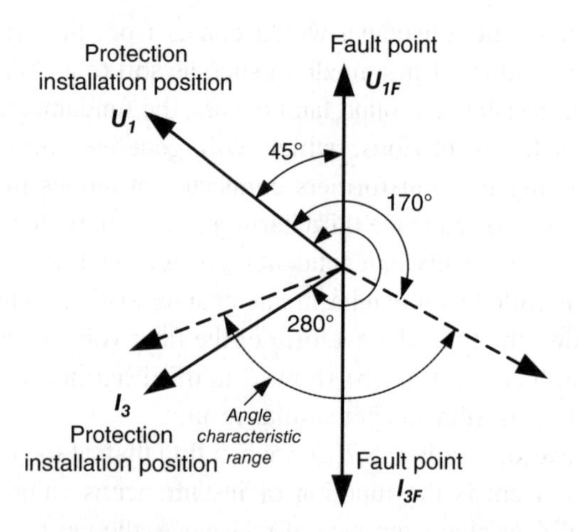

Figure 5.9 The phase relations between fault point and relay location

parameters (including the length of power lines, distributed capacitances, and equipment in parallel). So, according to different systems, we can determine the scope of the phase. At the measurement point in a power substation, if the phase angle difference between third harmonic current and fundamental voltage rises to a certain scope, and proportion of the amplitude of third harmonic and fundamental waves rises to a certain level, we can imagine that the harmonic characteristics for a high resistance ground fault appears, thus the relay should act to this harmonic. According to the examples shown in Figs 5.8 and 5.9, if the phase of the third harmonic current I_{3F} at the fault point lags 180° behind the fundamental voltage U_{1F}, at the measurement point in the power substation, considering the distributed capacitance and inductance, and the influence of these parameters' changes, the phase of the voltage at the measurement point U_1 and the third harmonic current I_3 will shift. Considering the line induction and capacitance in parallel, the analysis occurs next, as shown in Fig. 5.10.

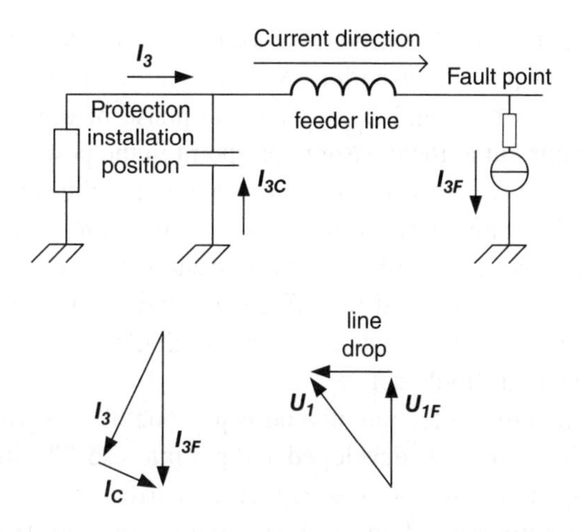

Figure 5.10 Phase shift analysis of fundamental voltage and third harmonic current

For voltage, because of the distributed inductions, the phase of the voltage U_1 at the measurement point will lead the voltage U_{1F} at the fault point. If the fault occurs near the measurement point, this angle can be ignored. If the fault occurs at the end of the line, the phase angle difference will be 15° between voltage at the measurement point and fault point, and 45° for the third harmonic. For current, analyzing the fault current in the fault circuit and taking fault point as a current source, because of the existence of the capacitive load in parallel (such as distributed capacitance and capacitance compensation device), the fault current at the measurement point (mainly comes from the resistive current at the neutral point) plus the capacitive current injected by distributed capacitance or capacitive load are equal to fault current at the fault point. As Fig. 5.10 shows, the phase of I_3 at the measurement point will lag behind the current at the fault point I_{3F}. Because of parallel capacitor, the third harmonic current I_3 at the measurement point lags 45° behind I_{3F}, the third harmonic current at the fault point. Considering the 45° of voltage phase shift, the third harmonic voltage at the relay location will lag 180 + 45 + 45 = 270° behind the fundamental voltage. Considering the capacitor's switch-on and the influence of branch load, we can imagine the third harmonic phase characteristics of a high resistance fault meet when the phase angle difference between fundamental voltage and third harmonic current at the measurement point rise to 170–280°.

In order to construct single-phase-to-ground fault protection based on the amplitude and phase of the third harmonic current, the operation criteria are:

$$\begin{cases} \dfrac{I_3}{I_1} > R_I \\ \varphi_{set2} > \varphi_{U1} - \varphi_{I3} > \varphi_{set1} \end{cases} \tag{5.3}$$

In which, I_3, I_1 are the third harmonic current and fundamental current at the measurement point in power substation, R_I is the ratio value, $\varphi_{set1}, \varphi_{set2}$ are setting values with units of degrees, and $\varphi_{U1}, \varphi_{I3}$ are the phase of the fundamental voltage and third harmonic current at the measurement point in the power substation.

Based on this analysis, considering the capacitors, the setting value can be estimated from the figure and previous analysis, we can set: $R_I = 10\%$; $\varphi_{set1} = 170°$; and $\varphi_{set2} = 280°$. Once the phase of third harmonic current increment lags 170–180° behind the fundamental voltage phase and the effective value of third harmonic relative to fundamental wave rises to 10%, we can imagine that the harmonic characteristics appear and the protection should act.

We need to note that this detection method is just one of the earliest high resistance fault detection methods that was developed and put into use. The high resistance fault detection has its specificity: the premise is that the current is not too high and causes little harm to the system, but is hard to detect, while the security risk still exits. So, unlike other relays, high resistance ground fault detections will not trip the circuit breaker generally. And the detection result is often evaluated by the probability of reliable and effective detection. So at present, in the field of high resistance ground fault detections, many methods are put forward.

5.3 Grounding Protection in the System with Neutral Isolated

The magnitude of zero sequence current is small when a single-phase-to-ground fault occurs in the system with neutral isolated, because the impedance (capacitance) of the fault circuit is large and the fault characteristics are not as distinct as those in a system with neutral earthed directly. For a two-phase-to-ground fault, phase-to-phase over current protection will take action to isolate the fault line. But for the single-phase-to-ground fault, the protection is only needed to issue an alarm signal selectively, not a trip signal generally.

5.3.1 Characteristics of Single-Phase-to-Ground Faults in Systems with Neutral Isolated

As shown in Fig. 5.11 in the simplest network wiring, under normal operation conditions, three phases have the same capacitances to ground. For each phase, there is current flowing through the capacitance to the ground and leading the voltage by 90° under the excitation of phase voltage. And the sum of three-phase currents is equal to zero. Assuming that a single-phase-to-ground fault occurs in phase A, the phase A voltage to ground becomes zero and the capacitance to ground is short-circuited as the other two phase voltages to ground rise to $\sqrt{3}$ times as high as phase-to-ground voltage and the current in the capacitance flowing to ground correspondingly rises to $\sqrt{3}$ times. The vector relationship is shown in Fig. 5.12. When a single-phase-to-ground

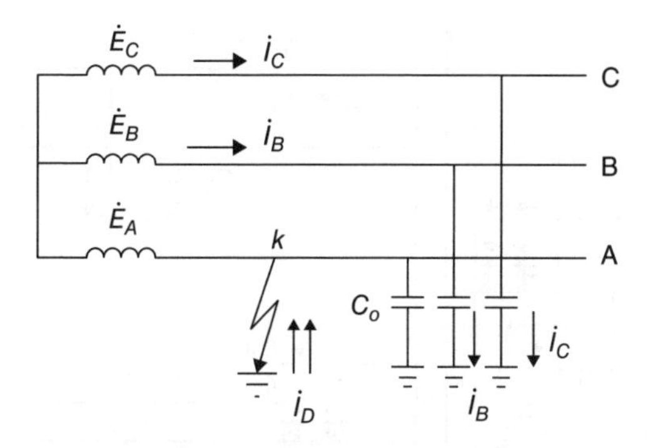

Figure 5.11 The schematic of simple network

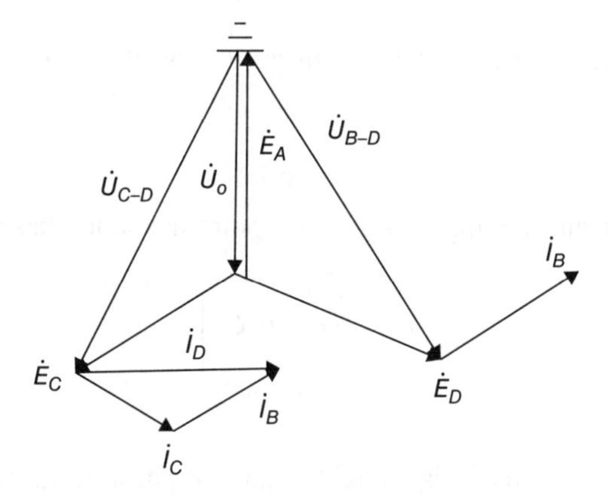

Figure 5.12 Vectors when phase A is grounded

fault occurs, three-phase load currents and line voltages are still symmetrical, so only the changes of the relationship to ground are analyzed.

After phase A is grounded, each phase voltage to ground is

$$\left.\begin{aligned}\dot{U}_{A-D} &= 0\\ \dot{U}_{B-D} &= \dot{E}_B - \dot{E}_A = \sqrt{3}\,\dot{E}_A\,e^{-j150^o}\\ \dot{U}_{C-D} &= \dot{E}_C - \dot{E}_A = \sqrt{3}\,\dot{E}_A\,e^{j150^o}\end{aligned}\right\} \tag{5.4}$$

The zero sequence voltage at the fault point k is

$$\dot{U}_{0k} = \frac{1}{3}\left(\dot{U}_{A-D} + \dot{U}_{B-D} + \dot{U}_{C-D}\right) = -\dot{E}_A \tag{5.5}$$

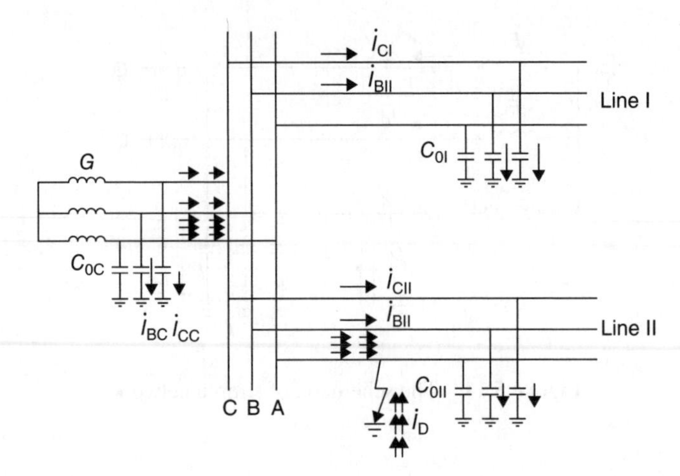

Figure 5.13 Capacitance current distribution diagram in three-phase system when single phase is grounded

The capacitance current flowing into the fault point in sound phases is that

$$\left.\begin{aligned} \dot{I}_B &= \dot{U}_{B-D}\, j\omega C_0 \\ \dot{I}_C &= \dot{U}_{C-D}\, j\omega C_0 \end{aligned}\right\} \tag{5.6}$$

RMS $I_B = I_C = \sqrt{3}U_\varphi \omega C_0$, as U_φ is the RMS value of phase-to-ground voltage.

At this time, current flowing from the grounding point is $\dot{I}_D = \dot{I}_B + \dot{I}_C$, and, as can be seen in Fig. 5.12, its RMS value is $I_D = 3U\varphi\omega c_0$, which is the arithmetical sum of three-phase capacitance current to ground under normal operation.

When there are generators and multiple lines in the system, as shown in Fig. 5.13, each generator and each line has a capacitance to ground, supposing lumped capacitors C_{0G}, C_{0I}, C_{0II}, and so on, describe them. When line IIA is grounded, if the voltage drop on the line impedance caused by load current and capacitor current is negligible, the voltage of phase A to ground is zero in the whole system, and the current flowing from phase A capacitance to ground of each element is also zero. Meanwhile, phase B and phase C to ground voltages and currents flowing from phase B and C capacitances to ground rise to $\sqrt{3}$ times, which can still be described by Equations (5.4)–(5.6). In this case, distribution of capacitance current is described by "→" in Fig. 5.13.

As shown in Fig. 5.13, in the healthy line I, phase A current is zero, and phase B and C currents are the current flowing through their own capacitances. Therefore, the zero sequence current at the start terminal of the line is

$$3\dot{I}_{0I} = \dot{I}_{BI} + \dot{I}_{CI} \tag{5.7}$$

Referring to Fig. 5.7, its RMS value is

$$3I_{0I} = 3U_{\varphi}\omega C_{0I} \tag{5.8}$$

That is, the zero sequence current is the current flowing through the capacitance of line I, and the direction of capacitive reactive power is from the bus to the line.

When there are multiple lines in the grid, this conclusion applies to each healthy line.

In generator G, firstly it has its own capacitance current to ground $\dot{I}_{BG}$ and $\dot{I}_{CG}$ in phase B and C, but it also generates another capacitance current, which causes all capacitance current from the fault point to flow back through phase A, while in phase B and C capacitance current to ground flows out to the same phase in each line. At this time, the zero sequence current at the generator's terminal is still the sum of current in three phases. As shown in the figure, because capacitance current flows into phase A and flows out to phase B and C in each line, which offset each other after summing, there's only the generator's own capacitance current. So,

$$3\dot{I}_{0G} = \dot{I}_{BG} + \dot{I}_{CG} \tag{5.9}$$

In which $3I_{0G} = 3U_{\varphi}\omega C_{0G}$. That is, the zero sequence current is the generator's own capacitance current, the direction of reactive power in the capacitance is from the bus to the generator, the same as the healthy line.

Now let's have a look at the fault line II. In phase B and C, the same as the healthy line, capacitance current $\dot{I}_{BII}$ and $\dot{I}_{CII}$ is flowing through it, while the difference is that the sum of capacitance current in phase B and C of the whole system is flowing back through the grounding point. The value is

$$\dot{I}_D = \left(\dot{I}_{BI} + \dot{I}_{CI}\right) + \left(\dot{I}_{BII} + \dot{I}_{CII}\right) + \left(\dot{I}_{BG} + \dot{I}_{CG}\right) \tag{5.10}$$

that is

$$I_D = 3U_{\varphi}w\left(C_{0I} + C_{0II} + C_{0G}\right) = 3U_{\varphi}wC_{0\Sigma} \tag{5.11}$$

In the formula $C_{0\Sigma}$ is the sum of all phase capacitance to ground of the whole system. This current is flowing through phase A. Therefore, the current flowing out of phase A can be described as $\dot{I}_{AII} = -\dot{I}_D$, so the zero sequence current flowing through the start terminal of line II is

$$3\dot{I}_{0II} = \dot{I}_{AII} + \dot{I}_{BII} + \dot{I}_{CII} = -\left(\dot{I}_{BI} + \dot{I}_{CI} + \dot{I}_{BG} + \dot{I}_{CG}\right) \tag{5.12}$$

That is

$$3I_{0II} = 3U_{\varphi}w\left(C_{0\Sigma} - C_{0II}\right) \tag{5.13}$$

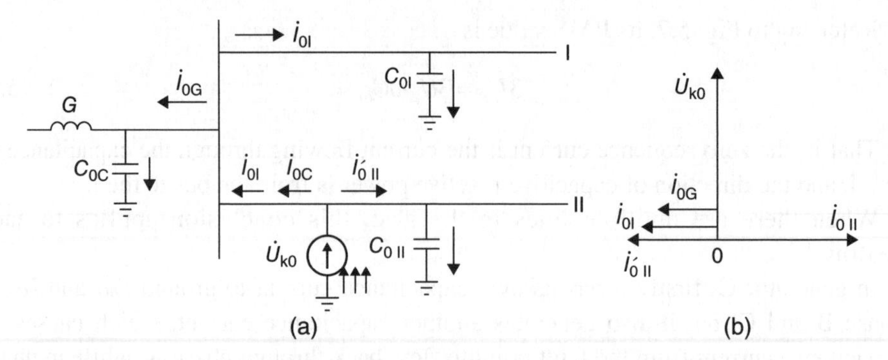

Figure 5.14 Zero sequence circuit (corresponding Fig. 5.13) and vectors

Thus, zero sequence current flowing from the fault line to the bus numerically equals the sum of capacitance current to ground in healthy elements of the whole system (except for the fault line itself). Its direction of the capacitive reactive power is from the line to the bus, which is opposite to the healthy line exactly.

According to the analysis result here, zero sequence equivalent network for a single-phase-to-ground fault can be made. As shown in Fig 5.14(a), the fault point has a zero sequence voltage $\dot{U}_{k0}$, while the loop of zero sequence current is constituted by capacitance between each element and ground. The zero sequence impedance of a transmission line can be negligible because it is much smaller than the capacitor's. A zero sequence current in an isolated neutral grid has exactly the same capacitance current in each element, the vectors' relationships of which are shown in Fig. 5.14(b). It's totally different from a solidly grounded neutral system.

For a single-phase-to-ground fault in the grid with neutral isolated, using the analysis in Fig. 5.13, a clear physical concept can be given out, while the calculation will be complex and inconvenient to use. But according to the analysis method in this figure, and after getting the equivalent zero sequence circuit shown in Fig. 5.14, it is convenient to calculate the value and distribution of zero sequence current. Concluding those analysis results previously, these following conclusions are drawn:

- When a single-phase-to-ground fault occurs, zero sequence voltage appears in the whole system.
- There is zero sequence current in healthy elements, the value of which equals its own capacitance current to ground. The actual direction of capacitive reactive power is from the bus to the line.
- In the fault line, zero sequence current is the sum of capacitance current to ground in all healthy elements of the whole system, the value of which is generally large. The actual direction of capacitive reactive power is from the line to the bus.

Those characteristics and differences will be the basis of the protection method.

5.3.2 *Single-Phase-to-Ground Protection in Grids with Neutral Isolated*

On the basis of the concrete network wiring, single-phase-to-ground protection can be constructed in the following ways.

5.3.2.1 Insulation Monitoring Device

At buses in power plants or power substations, single-phase-to-ground monitoring devices are generally installed, which will issue alarming signals after time delays based on zero sequence voltage generated by a fault.

As long as a single-phase-to-ground fault occurs, zero sequence voltage appears on all the buses in power plants and substations at the same voltage level. Therefore, signals issued by this method are not selective. If the fault line needs to be selected, it's required that the operation staff disconnect each line in turn, following an automatic reclosing to energize the disconnected line. When a line is disconnected, the signal of zero sequence voltage disappears, which indicates the fault is on this line. This is the so-called switching line method.

5.3.2.2 Zero Sequence Current Protection

Utilizing the characteristic that zero sequence current in the fault line is larger than that in the healthy line, the protection can issue selective signals or trip signals.

This kind of protection is generally utilized on the lines that have a prerequisite to install zero sequence current transformers (such as cable lines or overhead power lines connecting with cables). On the other hand, when single-phase-to-ground fault current is large, which can overcome the influence of unbalanced current at zero sequence current transducer, the protection devices can be installed in the zero sequence circuit formed by three-phase current transducers.

According to the analysis in Fig. 5.14, when a single-phase-to-ground fault occurs, the zero sequence current in the healthy line is its own current flowing through capacitances. Therefore, to ensure the selectivity of the protection, the set current of the protection device I_{act} should be larger than the capacitance current in this line. Thus,

$$I_{act} = K_{rel} 3U_{\varphi} wC_0 \tag{5.14}$$

In the formula, C_0 is the capacitance between ground and the protected line.

5.3.2.3 Zero Sequence Power Direction Protection

Utilizing the characteristic that zero sequence power has different directions for the fault line and the healthy line, the protection can issue selective signals or trip signals. This kind of protection is generally utilized in grids, where zero sequence current protection cannot meet the requirements of sensitivity or the wiring is too complex.

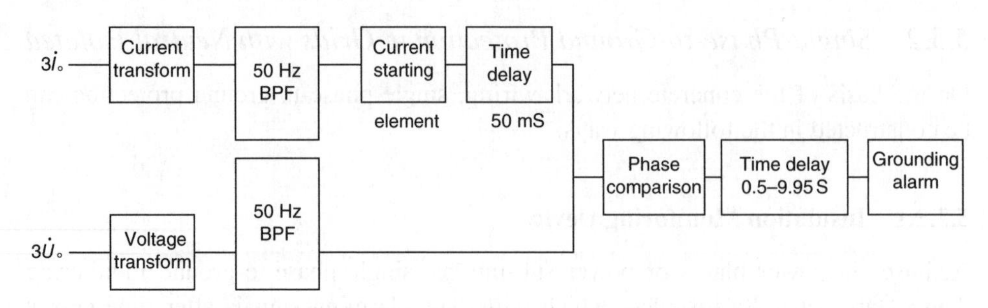

Figure 5.15 Construction diagram of zero sequence current direction protection

In order to improve the reliability and sensitivity of zero sequence directional protection, it can be considered only when a single-phase-to-ground fault occurs, the zero sequence current elements acts and after a delay of 50~100 ms there is an open phase comparison circuit in the directional element, as shown in Fig. 5.15.

The startup current of a zero sequence current device is set by the unbalance current in the zero sequence current transformer when a phase-to-phase fault occurs, which is independent from the current flowing through the capacitance in the protected element itself. It not only simplifies the setting calculations, but also improves the protection sensitivity greatly. The sensitive angle of the zero sequence directional element can be chosen as $\varphi_{sen\bullet\max} = 90^\circ$, which is the most sensitive when $3\dot{U}_0$ leading $3\dot{I}_0$ by 90°. The range of action is $\varphi_{sen\bullet\max} \pm (80^\circ \sim 90^\circ)$.

The feature of the scheme utilizing zero sequence current elements to control phase comparison circuit of zero sequence directional element is:

- The directional element works only when a grounding fault occurs, which improves the reliability.
- It is not affected by the unbalanced output from zero sequence voltage and zero sequence current filter under normal operating condition or under phase-to-phase faults.
- The phase comparison circuit starts to work after current elements start with a time delay of 50~100 ms, which can effectively prevent the effect caused by single-phase-to-ground transient processes on directional element.
- When the fault occurs outside of the protected zone, the current in protection is the current flowing through the capacitance in the protected element itself, and the direction element can block reliably.
- It cannot reflect an instantaneous grounding fault.

5.4 Grounding Protection in the System with Neutral Grounded Through an Arc Suppression Coil

An arc suppression coil installed at the neutral point is to reduce the grounding capacitance current when a single-phase-to-ground fault occurs, so the arc will not reignite. This also reduces the damage further caused by the ground fault. Meanwhile

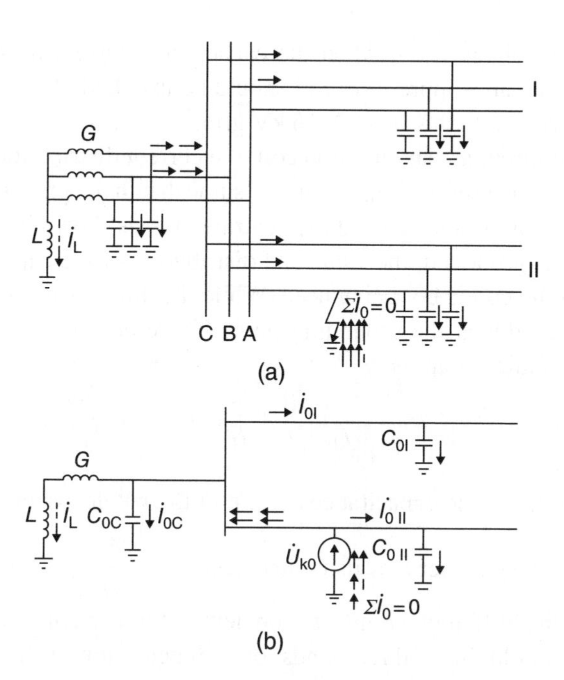

Figure 5.16 Current distribution when a single-phase-to-ground fault occurs in the grid with neutral grounded through arc suppression coil: (a) three-phase system, (b) zero sequence equivalent circuit

it also makes the selective grounding protection more difficult to implement, even the protection is only needed to issue an alarm signal.

5.4.1 Characteristics of Single-Phase-to-Ground Faults in Systems with Neutral Grounded through an Arc Suppression Coil

5.4.1.1 Steady-State Analysis of a Single-Phase-to-Ground Fault

Based on the analysis in the last section, when a single-phase-to-ground fault occurs in the system with neutral ungrounded, the capacitance current of the whole system flows through the neutral point. If this current is large, an arc will ignite at the grounding point and lead to arc overvoltage and the voltage to ground in the healthy line will increase. Therefore, the insulation may be damaged to cause two or more points to be grounded, even causing a blackout. In order to solve this problem, an inductance is usually connected at the neutral point, as shown in Fig. 5.16. So that when a single-phase-to-ground fault occurs, an inductive current component flows through the grounding point, which can offset the capacitance current in the original system and reduce the current flowing through the fault point. The inductance is called an arc suppression coil.

In grids at each voltage level, when the capacitance current in the whole system exceeds the value, an arc suppression coil should be installed: 30 A in a 3~6 kV grid, 20 A in a 10 kV grid, and 10 A in a 22~66 kV grid.

After the utilization of arc suppression coil, the current distribution changes greatly when a single-phase-to-ground fault occurs. Assume that in the grid shown in Fig. 5.16, an arc suppression coil is connected at the neutral point, as shown in Fig. 5.16(a), after phase A of line II is grounded, the value and distribution of the capacitance current is the same as when the coil is not being used. While the difference is that an inductance current component adds at the grounding point. Therefore, the total current flowing back from the grounding point is

$$\dot{I}_D = \dot{I}_L + \dot{I}_{C\Sigma} \tag{5.15}$$

In this formula, $\dot{I}_{C\Sigma}$ is the capacitance current of the whole system. $\dot{I}_L$ is the current in the arc suppression coil, set L as its inductance, $\dot{I}_L = \frac{-\dot{E}_A}{jwL}$.

According to the different compensation levels for capacitance current, an arc suppression coil could have three kinds of compensation: full, under, and over compensation.

- Full compensation makes $I_L = I_{C\Sigma}$, and current at the grounding point is approximately 0. From the point of view of avoiding arc overvoltage, it's the best way. However, when full compensation is used, $wL = \frac{1}{3wC_\Sigma}$, which is exactly the condition of 50 Hz AC series resonance between inductance L and capacitance. This will lead to a high rise in voltage to ground at the neutral point, so this way is rarely utilized.
- Under compensation makes $I_L < I_{C\Sigma}$, and the grounding current after the compensation is still capacitive. In this way, the problem is still unavoidable, because when the operation mode of the system changes, such as some element is out of service or isolated for a fault, the capacitance current will reduce. Then it is possible that I_L equals $I_{C\Sigma}$, leading to overvoltage again. Therefore, under compensation is also barely utilized.
- Over compensation makes $I_L > I_{C\Sigma}$, and the residual current after the compensation is inductive. And overvoltage caused by AC series resonance will not occur, so this approach is widely utilized in practice.
- The degree to which I_L is larger than $I_{C\Sigma}$ is expressed by the over-compensation degree P. The relationship is

$$P = \frac{I_L - I_{C\Sigma}}{I_{C\Sigma}} \tag{5.16}$$

Usually P is selected as percentages from 5 to 10%, and not more than 10%.

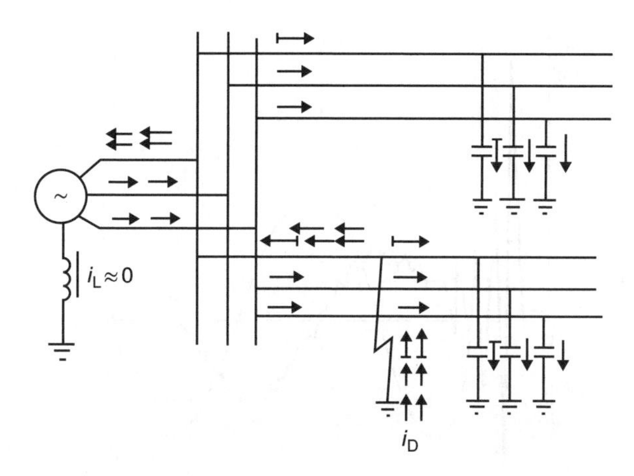

Figure 5.17 Transient current distribution under single-phase-to-ground fault

5.4.1.2 Transient Analysis of a Single-Phase-to-Ground Fault

When a single-phase-to-ground fault occurs, the transient component of the grounded capacitance current is possibly much higher than the steady-state current.

Generally, grounding faults caused by broken-down insulation always occur at the moment when the phase voltage is close to maximum. Therefore, the transient capacitance current could be regarded as the sum of the following two parts (see Fig. 5.17).

- The discharging capacitance current caused by the reduction of the fault-phase voltage, shown as $\mapsto$ in the figure. This current flows from the bus to the fault-point. The discharge current reduces rapidly and its frequency could reach thousands of Hz, which depends on the parameters in the grid (values of R and L), the location of the fault point and the fault resistance.
- The charging capacitance current caused by the rising of the sound phase voltage, shown as a $\rightarrow$ in the figure. Because the inductance in the current loop is large, the charging capacitance current damps slowly and the oscillation frequency is also low (just hundreds Hz). The waveform of the transient capacitance current at the fault point is shown as Fig. 5.18.

In an arc suppression coil grounded grid, the maximum of the transient inductance current occurs when the grounding fault occurs at the moment the phase voltage passing through 0, while $i_L \approx 0$ if the fault occurs when the phase voltage is close to the maximum. Therefore, the transient capacitance current is much higher than the transient inductance current. So, in the same grid, whether the neutral point is isolated or grounded through an arc suppression coil, the transient process is approximately the same: when the fault occurs at the time the phase voltage approaches the maximum.

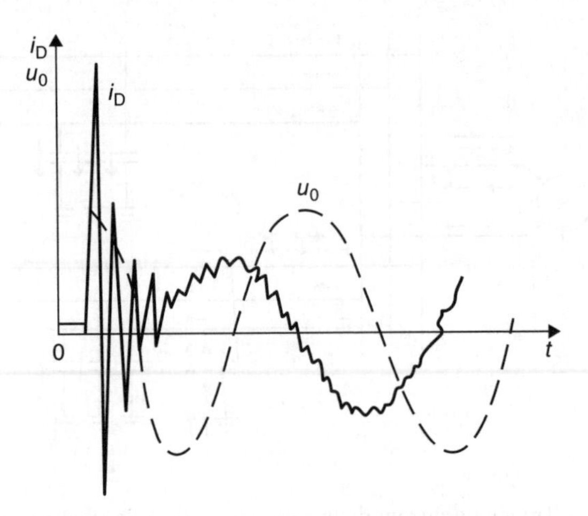

Figure 5.18 Transient single-phase-to-ground fault current waveform

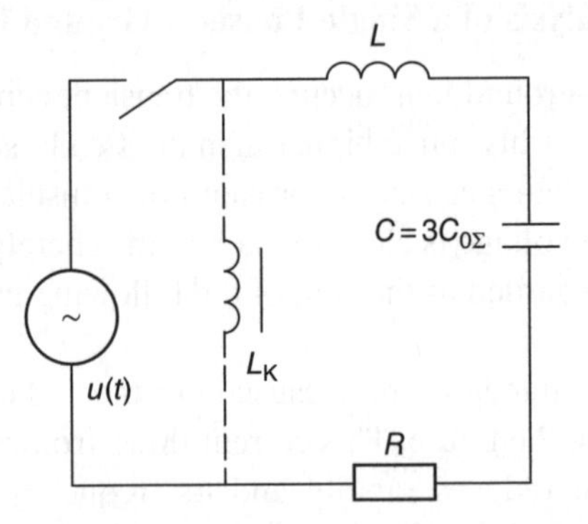

Figure 5.19 Equivalent circuit to analyze the transient process

During the transient process, an equivalent circuit can be used to estimate the grounding capacitance current component as shown in Fig. 5.19. The figure shows the distributed parameters R, L, C, and the lumped inductor of the arc suppression coil $L_K \geq L$. L_K doesn't affect the calculation of the capacitance current component actually, so it can be negligible. The resistance R, which affects the damp of the free oscillation, should be the whole resistance of the loop that the grounding current flows through, including the resistance of the lines, earth, and the fault resistance.

Neglecting L_K, the analysis of the transient capacitance current is to analyze the transient process of a series circuit formed by R, L, and C, which is suddenly excited by a zero sequence voltage $u(t) = U_m \cos wt$. Accordingly, the change tendency of the current flowing through the fault point mainly depends on the relationship of the circuit parameters R, L, and C. When $R < 2\sqrt{\frac{L}{C}}$, the transient process of the current has a periodical decay characteristic; while when $R > 2\sqrt{\frac{L}{C}}$, the current decays non-periodically and tends to a steady-state value.

For overhead lines, generally L is large and C is small, $R < 2\sqrt{\frac{L}{C}}$, therefore the fault point current has a rapid periodical decay characteristic. Based on the analysis results and measurements, free oscillation frequency is in the range of 300~1500 Hz. For the cables, L is small and C is large, therefore in comparison with overhead lines, the process time is extremely short and free oscillation frequency is high, in the range of 1500~3000 Hz.

5.4.2 Single-Phase-to-Ground Protection in Systems with Neutral Grounded through an Arc Suppression Coil

According to the analysis in Subsections 5.4.1 and 5.4.2, when a grounding fault occurs in the system with neutral grounded through an arc suppression coil, because of the compensation current from coil, the fault characteristic is not distinct enough, which makes it difficult to protect. In spite of this, engineers still try hard to solve this problem. The following are some cases of protection in the system with neutral grounded through an arc suppression coil.

5.4.2.1 Insulation Monitoring Devices

Using the zero sequence voltage after the fault occurs to judge the fault.

5.4.2.2 Using the Initial Half-Wave in the Transient Process at the Moment of the Grounding Fault to Construct the Protection

In China, the method using the first half-wave in the transient process to construct protection was proposed early in 1958. The basic principles are: (1) the amplitude of the initial half-wave grounding current in the transient process is large and (2) the initial half-wave zero sequence voltage and zero sequence current of the grounded line have opposite polarities. But because the peak value of the capacitance current is related to the instantaneous value of the phase voltage at the fault moment, it's difficult to ensure the protection devices act reliably.

5.4.2.3 Using the Active Power of the Coil to Construct the Protection

At the time when a fault occurs, the zero sequence current flowing through the sound line is only its own capacitance current, the phase of which leads to zero sequence voltage 90° and active power $P \approx 0$. When over-compensation is used, although the zero sequence current in sound lines is capacitive, the leading angle is less than 90°, because of the active power loss in the coil. The active power can be calculated by the following formula

$$p_{\circ} = \frac{1}{T}\int_{t}^{t+T} U_{\circ}(t) I_{\circ}(t) dt \tag{5.17}$$

In the formula, T is the period of the fundamental component, equal to 20 ms.

Assuming that the power loss in the coil is $p_{L\circ}$. The startup power of the protection is set as $P_{act} = 0.5P_L$. So the criterion of the fault line is $P_L > P_{act}$ and the healthy line is $P_L \leq (0.2\sim0.3)\, p_{act}$. This analysis is for the metallic fault. If the fault is through transition resistance, on one hand $U_{\circ}$ is smaller, and on the other hand the measured value is larger because of the power loss in the transition resistance. So, generally, the protection will act correctly.

5.5 Single-Phase-to-Ground Fault Feeder Selection Technology in a Power Distribution System with Neutral Non-Effectively Grounded

When a single-phase-to-ground fault occurs in a power distribution system with neutral ungrounded or grounded through a Peterson coil (non-effectively grounded or small current grounded), it is difficult to detect the faulted feeder due to a small short-circuit current. So far, there has been no ideal protection of single-phase-to-ground fault in such a power distribution system, even for warnings not to issue a signal to isolate the fault.

Neutral non-effectively grounded is widely used in 6~35 kV power distribution systems in China. The neutral in 6~10 kV power distribution system is ungrounded and the neutral in 35 kV power distribution systems is grounded through a Peterson coil. A single-phase-to-ground fault in the feeder occurs frequently due to low voltage and the short distance between conductors and conductors or between conductors and earth.

Single-phase-to-ground fault feeder selection is to determine whether there is a single-phase-to-ground fault and identify which feeder is grounded. The differences between fault feeder selection and above ground fault protection are:

- Single-phase-to-ground fault feeder selection focuses on single-phase-to-ground fault feeder in power distribution system.
- In general, the protection is for the protected component, and is based on electrical quantities of the protected component, while fault feeder selection is to select the fault feeder from many feeders and based on electrical quantities of many feeders.

Accordingly, for the special problem of single-phase-to-ground fault, fault feeder selection is an extended protection, which leads to a more and more flexible construction and implementation scheme than protection.

5.5.1 Comparison of Magnitude and Phase Based Single-Phase-to-Ground Fault Feeder Selection Methods

This method uses fault generated power-frequency information. When a single-phase-to-ground fault occurs in power distribution system with neutral ungrounded, there is zero-sequence current flowing through sound feeders, and the current of the sound feeder is equal to the current flowing through the capacitance of the feeder itself, while the current of the fault feeder is equal to the sum of the current flowing through the sound feeders. And the current of the sound feeder is reverse to the current of the fault feeder. The principle of the method is based on current magnitude and phase relations previously.

The detailed method can be implemented as follows:

- To detect zero-sequence voltage. If the zero-sequence voltage measured is bigger than the setting, it is determined that a single-phase-to-ground fault occurs;
- To calculate zero-sequence current in all feeders connecting to the same bus as the zero-sequence voltage bigger than the setting, and to select three feeders with the biggest current magnitude;
- To compare the current phase of three feeders selected, to determine the feeder with phase different from the other two feeders as the faulty feeder, and to determine that the fault occurs in the bus if the phases in three feeders are the same.

But for power distribution systems with neutral grounded through a Peterson coil, the zero-sequence current in the fault feeder is equal to the zero-sequence current in the sound feeder with the same flowing direction due to compensation by the Peterson coil. Accordingly, the method can't be used in a power distribution system with neutral grounded through a Peterson coil.

5.5.2 Characteristics of Single-Phase-to-Ground Fault Generated Current Traveling Waves

5.5.2.1 Generation and Features of Fault Generated Traveling Waves

As we all know, transmission and distribution power lines have the features of distributed parameters. But to analyze and calculate, it is always simplified to a lumped parameter, which is available for short power lines with low voltage. So far, fault analysis, fault detection, and protective relaying in power systems are all based on lumped parameter

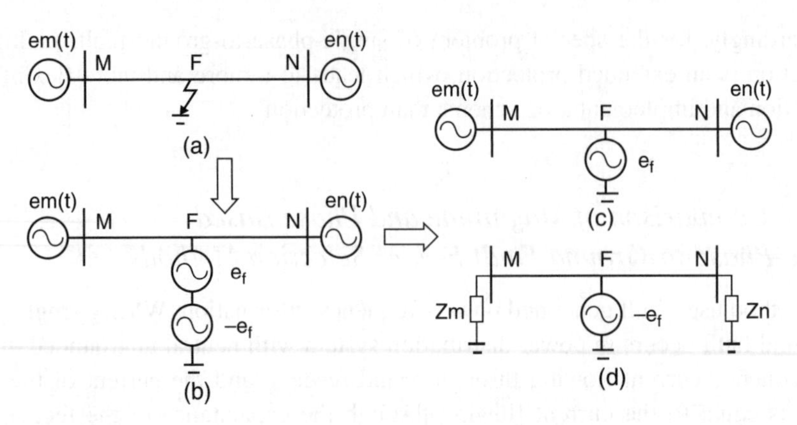

Figure 5.20 Post-fault power system and its decomposition. (a) Post-fault power system; (b) equivalent circuit of post-fault power system; (c) normal-operating network; and (d) fault superimposed network

circuit analysis. In fact, the accurate power line models have distributed parameters, and traveling waves are generated and propagated in the distributed parameter circuit.

5.5.2.2 Fault Generated Traveling Wave Source

According to the superimposed principle, the post-fault power system is equal to the faulty superimposed network superimposing a normal-operating network. In the faulty superimposed network, the superimposed source is the voltage source, and the value of superimposed source is the pre-fault voltage at the fault point. With the excitation of the superimposed voltage source, fault generated traveling waves are generated. Figure 5.20 shows the relation between a post-fault network, normal-operating network, and fault superimposed network. Figure 5.20(a) is post-fault power system, Fig. 5.20(b) is the equivalent circuit of post-fault power system in Fig. 5.20(a), Fig. 5.20(b) can be decomposed into the superimposition of normal-operating network in Fig. 5.20(c) and the fault superimposed network in Fig. 5.20(d). In the figure, e(t) stands for the superimposed voltage source in the fault superimposed network.

5.5.2.3 Propagation of Traveling Waves

When a fault occurs, the fault superimposed source at the fault node will transfer the voltage to other sound nodes under the excitation of the fault superimposed source. But there are capacitances and inductances in the distributed parameter circuit and the current through inductance and the voltage at capacitance can't change abruptly; time is needed to charge and discharge, which leads traveling waves to generate and

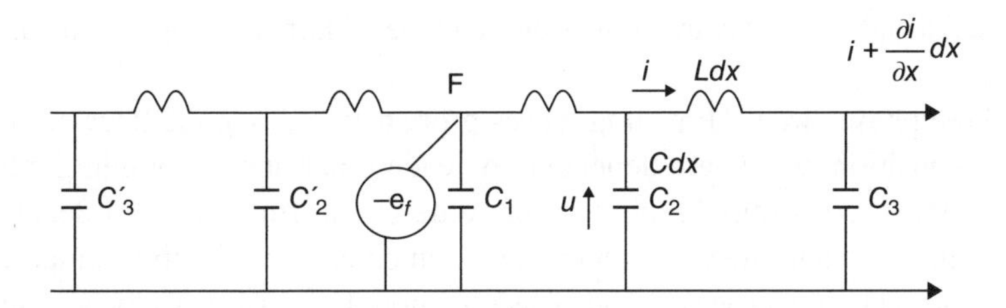

Figure 5.21 Single line equivalent circuit with distributed parameters and propagation of fault generated traveling waves

propagate. Obviously, the process of fault generated traveling waves is the same as for lightning, as shown in Fig. 5.21.

The relation between voltage traveling waves, current traveling waves, and the parameters of conductors can be represented as a wave equation:

$$\begin{cases} -\dfrac{\partial u}{\partial x} = L\dfrac{\partial i}{\partial t} \\ -\dfrac{\partial i}{\partial x} = C\dfrac{\partial u}{\partial t} \end{cases} \tag{5.18}$$

And it can be transformed to:

$$\begin{cases} \dfrac{\partial^2 u}{\partial x^2} = LC\dfrac{\partial^2 i}{\partial t^2} \\ \dfrac{\partial^2 i}{\partial x^2} = LC\dfrac{\partial^2 u}{\partial t^2} \end{cases} \tag{5.19}$$

In (5.18) and (5.19), L is the line inductance per unit length, the unit is H/km, C is the capacitance per unit length, the unit is F/km, u and i are voltage and current at the point with distance x away from fault point.

The general solution to Equation (5.19) is

$$\begin{cases} u = u_1\left(t - \dfrac{x}{v}\right) + u_2\left(t + \dfrac{x}{v}\right) \\ i = \dfrac{1}{Zc}\left[u_1\left(t - \dfrac{x}{v}\right) - u_2\left(t + \dfrac{x}{v}\right)\right] \end{cases} \tag{5.20}$$

In which $Zc = \sqrt{\dfrac{L}{C}}$ is wave impedance, $v = \dfrac{1}{\sqrt{LC}}$ is wave velocity, and $u_1\left(t - \dfrac{x}{v}\right)$ is forward traveling waves or positive direction traveling waves, which has the physical meaning that the waves travel away from the fault point along the positive direction with an increase in time, and $u_2\left(t + \dfrac{x}{v}\right)$ is reverse traveling waves or backward traveling waves.

The special solution to wave equations can be obtained using the detailed fault conditions.

For a three phase circuit, traveling waves propagating along the lines are functions of time and position, which are dependent on each other due to coupling inductances and capacitances. Accordingly, phase-to-module transform is used to decouple. The decoupled module quantities are independent, which can be used to calculate and analyze wave propagation process. The phase-to-module transform always used is the Karenbell transform, in which phase-to-module transform matrix S and module-to-phase transform matrix S^{-1} are:

$$S=\begin{bmatrix}1 & 1 & 1\\ 1 & -2 & 1\\ 1 & 1 & -2\end{bmatrix};\quad S^{-1}=\frac{1}{3}\begin{bmatrix}1 & 1 & 1\\ 1 & -1 & 0\\ 1 & 0 & -1\end{bmatrix} \tag{5.21}$$

And traveling wave components U and I in the phase space can be transformed to module space:

$$\begin{aligned} U_{\alpha\beta 0} &= \mathrm{S}^{-1}U_{ABC} = \frac{1}{3}\begin{bmatrix}1 & 1 & 1\\ 1 & -1 & 0\\ 1 & 0 & -1\end{bmatrix}\begin{bmatrix}U_A\\ U_B\\ U_C\end{bmatrix} \\ I_{\alpha\beta 0} &= \mathrm{S}^{-1}I_{ABC} = \frac{1}{3}\begin{bmatrix}1 & 1 & 1\\ 1 & -1 & 0\\ 1 & 0 & -1\end{bmatrix}\begin{bmatrix}I_A\\ I_B\\ I_C\end{bmatrix} \end{aligned} \tag{5.22}$$

$$\begin{cases} I_0 = \dfrac{I_A + I_B + I_C}{3} \\ I_\alpha = \dfrac{I_A - I_B}{3} \\ I_\beta = \dfrac{I_A - I_C}{3} \end{cases} \tag{5.23}$$

$$\begin{cases} U_0 = \dfrac{U_A + U_B + U_C}{3} \\ U_\alpha = \dfrac{(U_A - U_B)/3}{3} \\ U_\beta = \dfrac{U_A - U_C}{3} \end{cases} \tag{5.24}$$

In Equations (5.23) and (5.24), $U_\alpha, I_\alpha, U_\beta, I_\beta$ are line module components, and U_0, I_0 are zero module components.

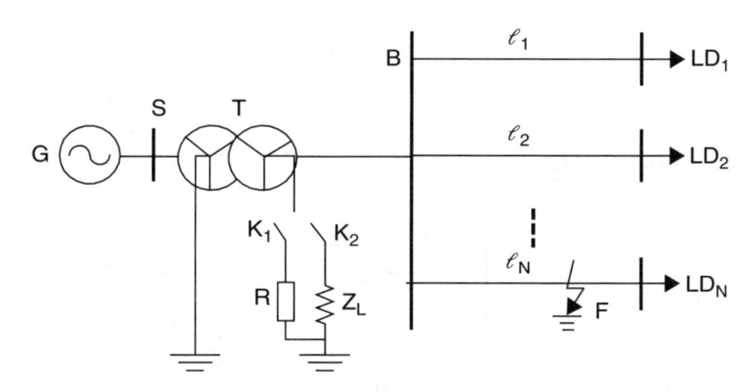

Figure 5.22 Power distribution system with one bus and N feeders

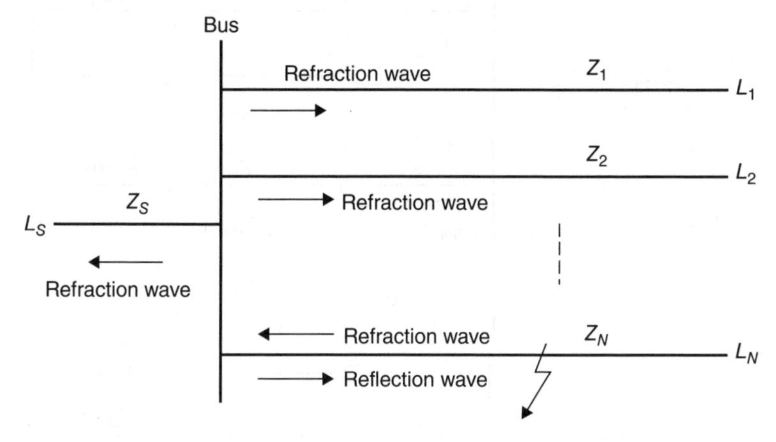

Figure 5.23 Reflection and refraction of incident traveling waves at the bus

5.5.2.4 Analysis of Single-Phase-to-Ground Fault Generated Current Traveling Waves Module Component

Figure 5.22 shows a power distribution system with one bus and N feeders. And the neutral in the power distribution system is non-effectively grounded. Through switches K_1 and K_2, neutral can be set to ungrounded, grounded through the Peterson coil, or grounded through resistance.

When phase A-to-ground fault occurs, traveling waves will be generated and propagated in the system under the excitation of fault superimposed source u_{FA}, which can be decomposed into three module voltages $u_{F\alpha}, u_{F\beta}, u_{F0}$. Traveling waves propagate from the fault point to the system along two sides of fault feeder. And the traveling wave arriving at the bus will be reflected and refracted by the bus. The initial traveling waves in the fault feeder are the superimposition of incident wave and reflected wave, while the initial traveling waves in the sound feeders are the refracted wave by the bus. The propagation process of traveling waves in the network can be demonstrated by Fig. 5.23.

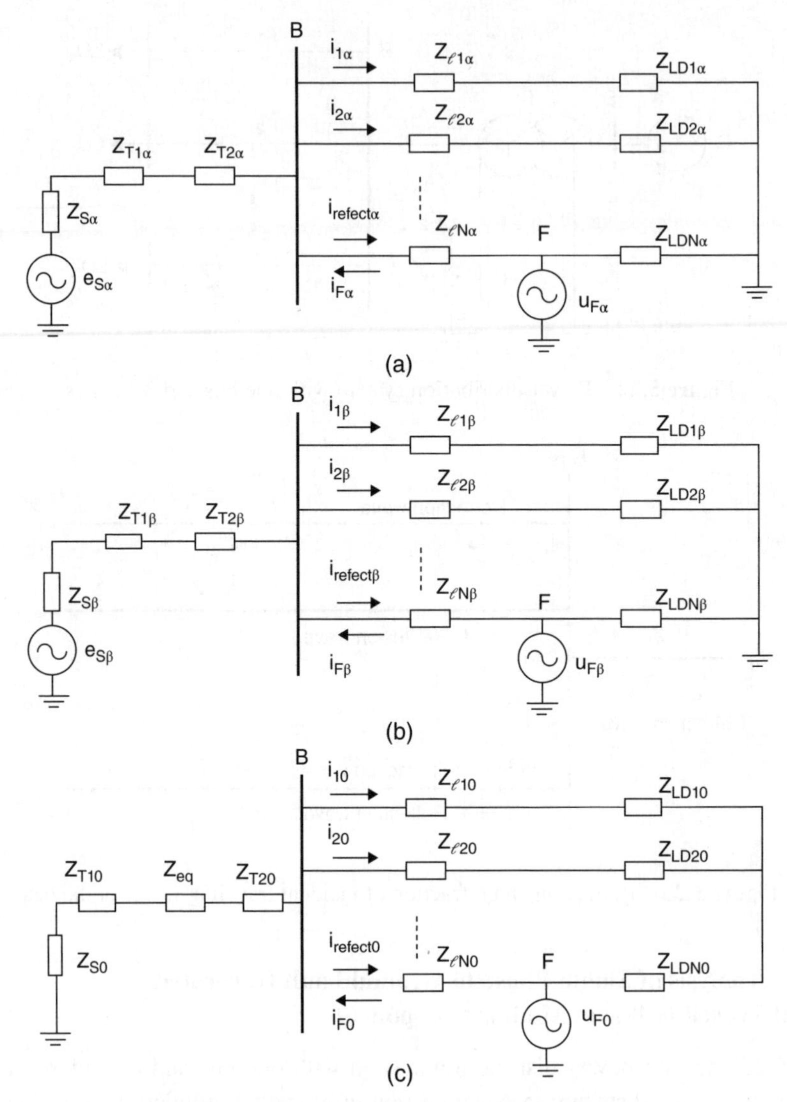

Figure 5.24 Module traveling waves equivalent circuit

Figure 5.24 shows the equivalent module circuit for the network in Fig. 5.22. Figure 5.24(c) shows a zero module equivalent circuit in which Z_{eq} is the equivalent wave impedance at the neutral point. When the neutral is ungrounded, Z_{eq} is infinite and when the neutral is grounded through a Peterson coil or resistance, Z_{eq} is the equivalent wave impedance of resistance or impedance at the neutral point.

In Fig. 5.24, $u_{F\alpha}$, $u_{F\beta}$, u_{F0}, module voltages of fault superimposed voltage sources at the fault point, respectively, are:

$$\begin{cases} u_{F\alpha} = \dfrac{u_{FA} - u_{FB}}{3} = -\dfrac{1}{3}u_{FA} \\ u_{F\beta} = \dfrac{u_{FA} - u_{FC}}{3} = -\dfrac{1}{3}u_{FA} \\ u_{F0} = \dfrac{u_{FA} + u_{FB} + u_{FC}}{3} = -\dfrac{1}{3}u_{FA} \end{cases} \tag{5.25}$$

$i_{F\alpha}$, $i_{F\beta}$, i_{F0}, the incident current traveling waves propagating from the fault point to the bus at a power substation, are

$$\begin{cases} i_{F\alpha} = -\dfrac{u_{F\alpha}}{Z_C} \\ i_{F\beta} = -\dfrac{u_{F\beta}}{Z_C} \\ i_{F0} = -\dfrac{u_{F0}}{Z_C} \end{cases} \tag{5.26}$$

When a single-phase-to-ground fault occurs at F in Feeder l_N, the wave impedance of the reflected traveling waves at the bus is a parallel connection of all sound feeders and the transformer. And equivalent wave impedance Z_{BX} can be represented as the following equation, in which x can be any module of α, β, and zero.

$$Z_{Bx} = \frac{1}{\dfrac{1}{Z_{T2x}} + \sum_{k=1}^{N-1} \dfrac{1}{Z_{Tkx}}} \tag{5.27}$$

In Equation (5.27), Z_{T2x} is the equivalent wave impedance of transformer secondary side, and Z_{lkx} is the wave impedance of Feeder l_k (k=1~N – 1).

When a phase A-to-ground fault occurs, and the incident current traveling waves i_{Fx} (x means module quantity) arrive at the bus from the fault point due to reflection and refraction at the bus, the reflected wave $i_{refractx}$ and the refracted wave $i_{reflectx}$ are, respectively,

$$i_{refractx} = -\frac{2Z_{Bx}}{Z_{Bx} + Z_{Nx}} i_{Fx} \tag{5.28}$$

$$i_{reflectx} = \frac{Z_{Nx} - Z_{Bx}}{Z_{Bx} + Z_{Nx}} i_{Fx} \tag{5.29}$$

In which, i_{Fx} is the incident current traveling wave, Z_{Bx} is the equivalent wave impedance at the bus, and Z_{Nx} is the wave impedance of faulted feeder l_N.

The initial current traveling waves of faulted feeder N measured at the bus is the sum of the incident wave and the reflected wave,

$$i_{Nx} = -\frac{2Z_{Nx}}{Z_{Bx} + Z_{Nx}} i_{Fx} \tag{5.30}$$

If we assume the wave impedance of sound feeder l_k is Z_k, the initial current traveling wave of the sound feeder is the refraction wave at the bus, as

$$i_{kx} = \frac{2Z_{Bx}^2}{Z_{kx}\left(Z_{Nx} + Z_{Bx}\right)} i_{Fx}, \left(k = 1 \sim N-1\right) \tag{5.31}$$

Compare Equation (5.29) with (5.30) and it can be found that, because many feeders are connected to the bus in a power distribution substation, and the equivalent wave impedance of the bus is much smaller than the wave impedance of the feeder (i.e., ZBx << Zkx), the initial traveling wave magnitude of the faulted feeder is much higher than the wave of sound feeder with the reverse polarity while the initial traveling wave magnitudes of all sound feeders are close to equal and with the same polarity. Most importantly, this analysis shows that the initial traveling waves are independent from the method of neutral grounding, which presents a novel solution to fault feeder selection in a power distribution system with neutral non-effectively grounded.

5.5.3 Current Traveling Wave-Based Fault Feeder Selection Method

5.5.3.1 Single-Phase System

From above analysis, there are obvious different characteristics between initial traveling waves in the fault feeder and in sound feeders. Accordingly, the principle is based on initial current traveling waves. The method is based on initial traveling waves of all feeders and selects the fault feeder using magnitude and polarity difference between initial current traveling waves of sound feeders and the wave of the fault feeder.

The fault feeder selection criterion is that the feeder with the highest magnitude of initial traveling wave in all feeders and with the reverse polarity to other feeders is selected as the fault feeder, which can be represented as

$$\begin{cases} \left|I_f\right| > \left|I_k\right| & k \in Z \\ \operatorname{sgn}\left(I_f\right) & \neq \operatorname{sgn}\left(I_k\right) \end{cases} \tag{5.32}$$

In which I is the initial current traveling waves, f and k are feeder number, Z is the total number of feeders.

5.5.3.2 Three-Phase System

A power system is a three-phase system and phase traveling waves couple with each other, so the method used in the single-phase system can't be applied directly. With phase-to-module transform, module traveling waves are independent, and this method in a single-phase system can be used to select the fault feeder. Therefore, the method for a three-phase system is as follows:

- For the system with zero-sequence current transducers, zero module current traveling waves can be used directly to select the fault feeder.
- For the system with three phase current transducers, all module current traveling waves can be obtained based on phase-to-module transform, and module α and β current traveling waves can be used to select the fault feeder. Because of module α traveling waves comprise of phase A and Phase B information, module α current traveling waves can be used to select phase A and phase B grounded. Similarly, module β traveling waves can be used to select phase A and phase C grounded.

Whatever module α, β, γ, or zero current traveling waves are used to select the fault feeder, the criterion is the same as Equation (5.32).

In fact, it is also available using phase A and phase C current traveling waves to select the fault feeder, even based on single-phase current traveling waves.

5.6 Prevention of and Protection from Single-Phase-to-Ground Faults in Power Distribution Systems with Neutral Non-Effectively Grounded

The voltage in the power distribution system is low and the fault usually has an evolving process: it begins with transient flashover of insulation, proceeds to dielectric breakdown and discharge, and finally forms permanent insulation breakdown and becomes a fault. The field operation of traveling waves based single-phase-to-ground fault line selector has proved that not only does a fault have a process, but also has recorded traveling wave information during the fault forming process. This traveling wave information makes research of preventive protection possible. If we can detect signs before a permanent fault occurs, it is possible to prevent a fault and reduce accidents.

5.6.1 Basic Principle of Single-Phase-to-Ground Fault Prevention

On-line power equipment is under the rated AC voltage. If the power equipment's insulation is higher than applied voltage, the equipment can operate normally. Along with equipment aging or environment changing, insulation will decrease. With many kinds of events, a power system may generate overvoltage. While applied voltage is higher than insulation voltage instantly, a transient flash-over will occur in power equipment. At the flash-over point, traveling waves will be generated. No matter whether the events are transient flash-over, partial discharge, or insulation breakdown, the relevant equipment will generate traveling waves and such an event-generated traveling wave will propagate in the system along the power line. According to traveling wave propagation theory, at the wave impedance discontinuity point (e.g., a bus bar in a power substation), traveling waves will be reflected and refracted. The refracted traveling waves will penetrate other normal operation lines. Regardless of sources (power discharge, insulator flash-over, insulation breakdown, single-phase-to-ground fault), the initial traveling wave's characteristic is the same. Traveling waves in the line of the disturbance source have the biggest amplitude and opposite polarity to the other line's traveling waves.

However, traveling waves in a distribution line can be generated not only by insulation flash-over, partial discharge, insulation breakdown, and short circuit faults, but also by circuit breaker operation, lightning strike, and so on. Similarly, traveling waves in the line of the disturbance source have the biggest amplitude and opposite polarity to the other line's traveling waves. While a short-circuit fault happens in system, not only does it generate traveling waves at the fault point at inception, but typical power frequency variations will also change during the fault (voltage decreases, current increases, etc.). However, circuit breaker operation and lightning strike, on one hand, will not cause a fault. So the system can recover normal operation and power frequency electric parameters won't make a significant change after the event. On the other hand, circuit breaker operation and lightning strike in normal systems are accidental events, which mean the interval between two events may be long. But considering the process from insulation flash-over, partial discharge to insulation breakdown, although the process will not generate significant changes of power frequency electric quantities, the frequency of traveling wave generated will get higher and higher depending on the source.

Therefore, fault evolving process generated traveling waves are continuous, and the time interval between two waves will get shorter and shorter from the same source, while for other event-generated traveling waves, the interval is relatively random, discrete, and long.

5.6.2 *Single-Phase-to-Ground Fault Prevention Technology*

Single-phase-to-ground fault prevention technology mainly includes identification of traveling wave source and type. Traveling wave source identification is based on mutual relation of all possible traveling waves. Identification is based on the time interval of traveling waves generated by the same source.

Figure 5.25 shows a flow chart of single-phase-to-ground prevention. The technology acquires traveling waves in all distribution feeders in real-time, and determines the startup at the same time. The disturbance is detected in real time. If there is no disturbance detected, the acquired traveling waves will be covered by new data.

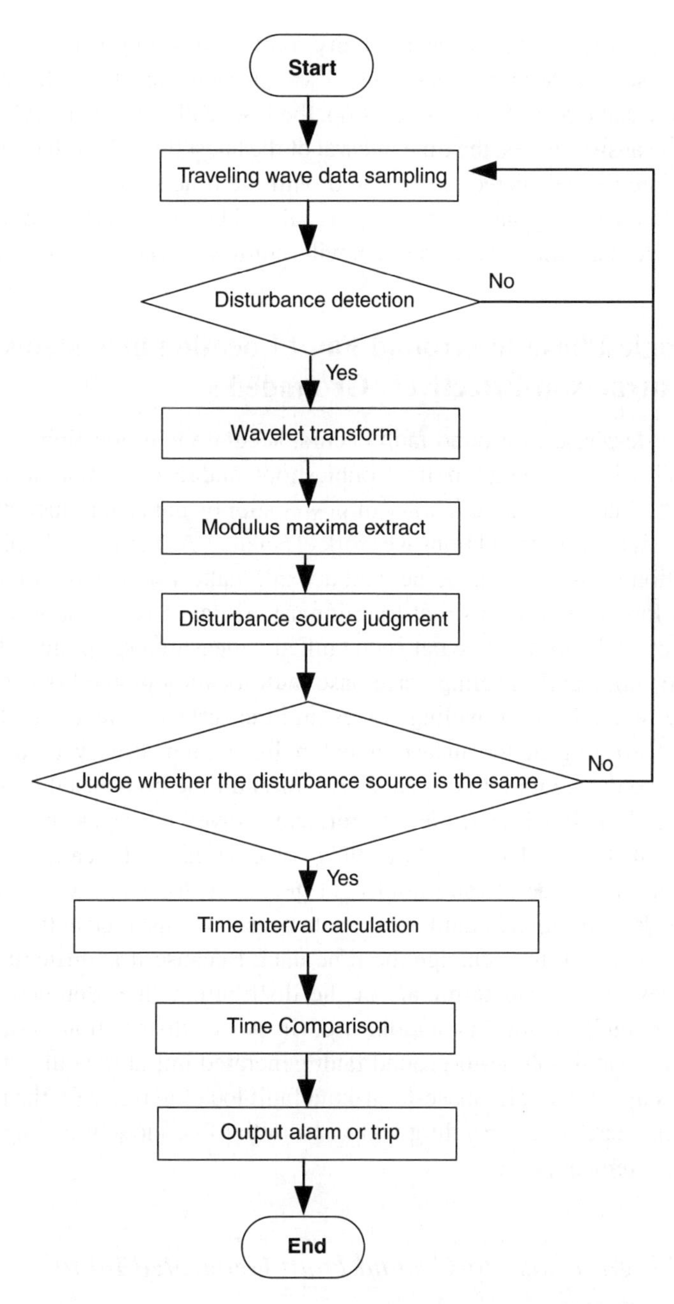

Figure 5.25 Flow chart of single-phase-to-ground fault prevention

If a disturbance is detected, the acquired traveling waves will be processed using wavelet transform. The maxim of wavelet transform of traveling waves is extracted. Then the amplitude and polarity of the maxim of wavelet transform of currents in all distribution feeders are compared with each other. The distribution feeder with the highest amplitude and the reverse polarity to other traveling waves is selected as the disturbance source. Next, it is determined whether the latest three disturbance sources are from the same one. If the answer is no, the flow will return to traveling wave acquisition. If the answer is yes, the time interval of the latest three disturbances is calculated. The time interval calculated is compared with the time setting. If the time interval is shorter than the setting, an alarm or trip signal will be issued. If the time interval calculated is longer than the setting, the flow will return to traveling wave acquisition.

5.7 Single-Phase-to-Ground Fault Location in Systems with Neutral Non-Effectively Grounded

After a single-phase-to-ground fault occurs, accurate fault location will help maintenance staff arrive at the fault point, troubleshoot, and recover the power supply, which will not only increase the reliability of power supply, but also reduce the workload of line patrol significantly and improve work efficiency. As far as single-phase-to-ground fault location in such a system, the fault doesn't make a short-circuit loop, so the fault current is low, which means that traditional impedance based fault location method can't calculate the impedance between fault and measurement point to locate the fault point. The single end traveling wave base fault location method is based on the time difference of the initial traveling waves and the reflected traveling waves from the fault point arriving at the measurement point, which is easy to use in the field. However, the distribution line is so short with multiple branches that the reflection is strong, which makes identifying the reflected traveling waves from the fault point very difficult. The double-ended traveling wave based fault location method is based on the time difference of the initial fault generated traveling waves arriving at two ends of the line. It requires fault location devices to be installed at two ends of the line and communication to exchange the time data. Because it is difficult to install fault location devices at load terminals of the distribution line, double-ended traveling wave based fault location is seldom used in power distribution systems. Based on reanalysis of single-phase-to-ground fault generated initial traveling waves, this section comes up with single-phase-to-ground fault location based on the time difference of the initial zero mode traveling wave and initial line mode traveling wave arriving at the measurement point.

5.7.1 Single-Phase-to-Ground Fault Generated Initial Traveling Waves

When phase A-to-ground fault through resistance occurs in the power system, according to superimposed theory, the faulted network can be divided into normal load and superimposed networks. Obviously, initial traveling waves are in the superimposed network. Assuming that power lines are transposed, the superimposed network can

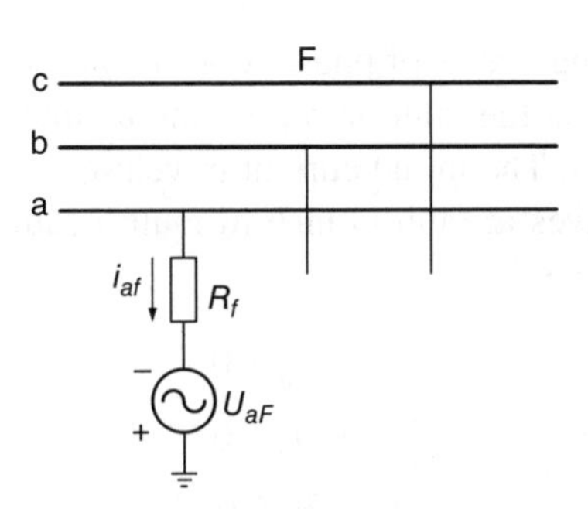

Figure 5.26 Boundary conditions of single-phase-to-ground fault

be divided into two parts at the fault point, one system side including transposed lines, the other faulted branches. For a system side, the voltage and current meet the independent relationships in module quantities. As Equation (5.33) shows:

$$\begin{cases} u_0 = Z_0 i_0 \\ u_1 = Z_1 i_1 \\ u_2 = Z_1 i_2 \end{cases} \tag{5.33}$$

In which, u_0, u_1, u_2 are module 0, module 1, and module 2 voltage traveling waves at fault point, respectively; i_0, i_1, i_2 are module 0, module 1, and module 2 current traveling waves of power line; and Z_0, Z_1, Z_2 are module 0, module 1, and module 2 wave impedance.

When a single-phase-to-ground fault (like phase a) occurs in the system, as shown in Fig. 5.26, the equation in faulted branch can be expressed by:

$$\begin{cases} u_{af} - R_f i_{af} + u_{aF} = 0 \\ I_{bf} = 0 \\ I_{cf} = 0 \end{cases} \tag{5.34}$$

In which, u_{af} is initial voltage phase A traveling wave at fault point t, i_{af} i_{bf} i_{cf} are phase A, phase B, and phase C faulted branch currents, respectively, u_{aF} is a phase A superimposed voltage source at the fault point, and R_f is fault resistance.

What's more, the voltage of system side at the fault point is the same as the voltage of fault branch at the fault point, as Equation (5.35) shows.

$$\begin{cases} u_{af} = u_a \\ u_{bf} = u_b \\ u_{cf} = u_c \end{cases} \tag{5.35}$$

u_{af}, u_a are the initial traveling waves of phase A at the fault point; u_{bf}, u_b are the initial traveling waves of phase B at the fault point; u_{cf}, and u_c are the initial traveling waves of phase C at the fault point. The initial current traveling wave of system side and the initial current traveling waves at fault branch in fault location should meet the KCL law, as Equation (5.36) shows:

$$\begin{cases} i_{af} + 2i_a = 0 \\ i_{bf} + 2i_b = 0 \\ i_{cf} + 2i_c = 0 \end{cases} \tag{5.36}$$

In which, i_{af}, i_{bf}, i_{cf} are the initial current traveling waves of phase A, B, and C in the fault branch. i_a, i_b, i_c are the initial current traveling waves of phase A, B, and C on the system side.

To simplify the analysis, we assume that the lines have been transposed fully and use the Karenbell transformation matrix as the phase-mode transform matrix. Therefore, the three independent modulus equation of system side can be transformed into the following equation:

$$\begin{cases} u_a + u_b + u_c = Z_0\left(i_a + i_b + i_c\right) \\ u_a - u_b = Z_1\left(i_a - i_b\right) \\ u_a - u_c = Z_1\left(i_a - i_c\right) \end{cases} \tag{5.37}$$

In which, u_a, u_b, u_c are the initial voltage traveling waves of phase A, B, and C; and i_a, i_b, i_c are the initial current traveling waves of phase A, B, and C in the lines.

After solving these equations, the initial voltage traveling waves in phase quantities at fault point and initial current traveling waves in the lines are:

$$\begin{cases} u_a = -\dfrac{Z_0 + 2Z_1}{Z_0 + 2Z_1 + 6R_f} u_{aF} \\ u_b = -\dfrac{Z_0 - Z_1}{Z_0 + 2Z_1 + 6R_f} u_{aF} \\ u_c = -\dfrac{Z_0 - Z_1}{Z_0 + 2Z_1 + 6R_f} u_{aF} \end{cases} \tag{5.38}$$

$$\begin{cases} i_a = -\dfrac{3}{Z_0 + 2Z_1 + 6R_f} u_{aF} \\ i_b = 0 \\ i_c = 0 \end{cases} \tag{5.39}$$

Because different module traveling waves have different propagation speeds in power lines, the initial phase traveling wave can't be obtained at the measurement point, only the initial module traveling waves. Therefore, according to the module traveling wave propagation speed, the initial module traveling waves obtained at the measurement point can be represented as

$$\begin{cases} u_0 = -\dfrac{Z_0}{Z_0 + 2Z_1 + 6R_f} u_{aF} \\ u_1 = -\dfrac{Z_1}{Z_0 + 2Z_1 + 6R_f} u_{aF} \\ u_2 = -\dfrac{Z_1}{Z_0 + 2Z_1 + 6R_f} u_{aF} \end{cases} \tag{5.40}$$

$$\begin{cases} i_0 = -\dfrac{1}{Z_0 + 2Z_1 + 6R_f} u_{aF} \\ i_1 = -\dfrac{1}{Z_0 + 2Z_1 + 6R_f} u_{aF} \\ i_2 = -\dfrac{1}{Z_0 + 2Z_1 + 6R_f} u_{aF} \end{cases} \tag{5.41}$$

For different module traveling waves with different propagation speeds, the higher speed line module traveling wave arrives at the measurement point earlier, and the lower speed zero module traveling wave arrives at the measurement point later, which can be represented as

$$\begin{cases} i_{apl} = -\dfrac{2}{Z_0 + 2Z_1 + 6R_f} u_{aF} \\ i_{bpl} = \dfrac{1}{Z_0 + 2Z_1 + 6R_f} u_{aF} \\ i_{cpl} = \dfrac{1}{Z_0 + 2Z_1 + 6R_f} u_{aF} \end{cases} \tag{5.42}$$

$$\begin{cases} i_{apz} = -\dfrac{1}{Z_0 + 2Z_1 + 6R_f} u_{aF} \\ i_{bpz} = -\dfrac{1}{Z_0 + 2Z_1 + 6R_f} u_{aF} \\ i_{cpz} = -\dfrac{1}{Z_0 + 2Z_1 + 6R_f} u_{aF} \end{cases} \tag{5.43}$$

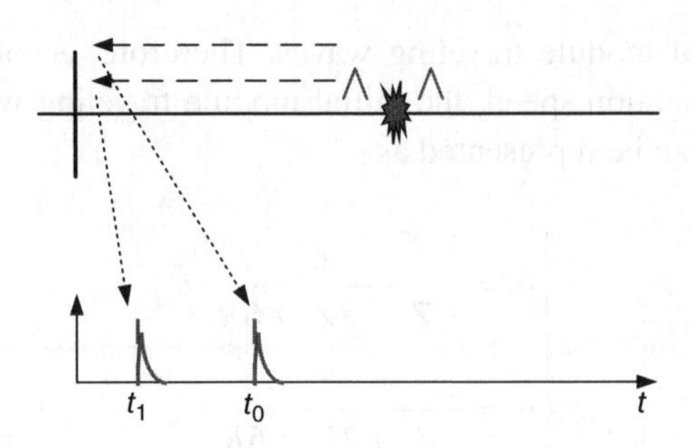

Figure 5.27 Basic principle of single-phase-to-ground fault location based on propagation speed of traveling waves

In which i_{apl}, i_{bpl}, i_{cpl} are initial line module current traveling waves of phase A, B, and C; and i_{apz}, i_{bpz}, i_{cpz} are initial zero module current traveling waves of phase A, B, and C.

Therefore, in power distribution systems, the initial traveling waves at phase current transducers are line module traveling waves, while the initial traveling wave at the zero sequence current transducer is a zero module traveling wave. Based on this, a novel single-phase-to-ground fault location method can be proposed in power distribution systems with neutral non-effectively grounded.

5.7.2 Single-Phase-to-Ground Fault Location Method Based on Propagation Speed of Traveling Waves

When a single-phase-to-ground fault occurs in the line as shown in Fig. 5.27, the fault generated traveling waves will propagate along the line from the fault point. And according to different module traveling waves, the speed of traveling waves will be different too. The speed of line module traveling wave's speed v_1 is faster than the speed of the zero module traveling wave v_0. For the same distance x from fault point to measurement point, line module traveling wave will arrive at the measurement point first at time t_1, while zero module traveling wave will arrive later at time t_0. Assuming that the fault time is t_r, then:

$$\begin{cases} x = v_1 (t_1 - t_r) \\ x = v_0 (t_0 - t_r) \end{cases} \tag{5.44}$$

Solve the equation and we can get

$$x = v_1 v_0 (t_1 - t_0) / (v_1 - v_0) \tag{5.45}$$

In which the propagation speed of traveling waves can be determined by line type and operation environment, the arrival time of traveling waves can be measured directly.

In the power distribution system equipped with zero sequence current transducers, the initial zero module traveling wave is the first monitored by the zero sequence current transducer. In the power distribution system without zero sequence current transducer, the initial zero module traveling wave can be the first wave of the sum of three phase current traveling waves measured by phase transducers. And the initial line module traveling wave can be the first wave measured by the fault or sound phase current transducers. Therefore, according to different current transducers configuration in the system, the method can have different implementation schemes.

5.8 Conclusion and Summary

Single-phase-to-ground fault is the main fault in power distribution system. According to different types of neutral grounding in power distribution systems, the fault characteristics and effects are different, which requires different solutions. The chapter presents different processing solutions to single-phase-to-ground fault in power distribution system including neutral effectively and neutral non-effectively grounded systems after analyzing the fault characteristics in details.

For a single-phase-to-ground fault in power distribution system with neutral effectively grounded, zero-sequence over current protection can detect the fault correctly. However, if the fault is high impedance fault, which occurs frequently in power distribution systems, the fault characteristics are not obvious and over current protection may not operate properly. The chapter presents the solution based on three harmonic current amplitude and phase, which paves the way to high impedance fault detection. The single-phase-to-ground fault generated zero sequence current is small in power distribution systems with neutral ungrounded, which requires the alarm signal issued by the protection and occasionally the trip signal. In order to compensate for the capacitive current when a single-phase-to-ground fault occurs, the Peterson coil is installed at neutral, which makes it more difficult to protect the single-phase-to-ground fault in such a system.

For single-phase-to-ground fault in power distribution systems with neutral non-effectively grounded, the chapter presents traveling wave based faulted feeder selection after analyzing fault generated traveling waves, which improves the sensitivity and reliability of protecting a single-phase-to-ground fault. And with recording of single-phase-to-ground fault generated traveling waves in the field, the chapter proposes fault prevention, which can forecast the fault based on pre-fault information, especially traveling waves. In order to accurately locate the fault to recover the power supply as fast as possible, traveling wave based single-phase-to-ground fault location is introduced in the chapter, which is based on the time difference between a single-phase-to-ground fault generated initial line module traveling wave and initial zero module traveling waves.

6

Practical Aspects of Fault Processing

Liu Jian and Zhang Xiaoqing

Abstract

The fault processing approach to coordinate the fault processing approaches of local, distributed, and centralized intelligence to make the best of them is put forward. The approach to plan the amount of various FTUs in DAS to meet the requirement of service reliability is described. The test methodologies to verify the properties of fault location, isolation, and service restoration of a fault processing system based on local, distributed, or centralized intelligence are proposed.

Keywords

fault processing procedure, comparison of various fault processing methodologies, coordination of various fault processing approaches, planning the amount of terminal units, feeder terminal unit (FTU), fault indicator, relay protection, hybrid mode of FTU and fault indicator, master injection testing methodology, secondary synchronous injection testing methodology, master and secondary synchronous injection testing methodology, direct short-circuit test

6.1 Introduction

Each of the fault processing approaches, no matter whether it is based on local, distributed, or centralized intelligence, has its advantages and disadvantages. In practice, we should scientifically plan the fault processing system to coordinate fault processing approaches with each other to make the best of them, which will be discussed in Section 6.2.

Fault Location and Service Restoration for Electrical Distribution Systems, First Edition. Jian Liu, Xinzhou Dong, Xingying Chen, Xiangqian Tong, Xiaoqing Zhang and Shiming Xu.

The simpler a system, the more reliable it is. From another point of view, the fewer FTUs, the more economical the system is. Section 6.3 covers FTU amount planning to meet the requirement of service reliability.

Section 6.4 describes the test methodologies to verify the property of fault location, isolation and service restoration of a fault processing system based on local, distributed, or centralized intelligence, some of which lead to service interruption, the others do not.

6.2 Coordination of Fault Processing Approaches

Combining the fault processing approaches of local, distributed, and centralized intelligence may improve the fault processing property.

6.2.1 Fault Processing Performance of Various Methodologies

The whole fault processing procedure consists of six steps as shown in Figure 6.1, in which clear control, fault location, fault isolation, service restoration, and return control can be carried out automatically within a short period. But repairs must be done by man-power, which may take a rather long time.

6.2.1.1 Clear Control

Clear control must be carried out by local intelligence. A circuit breaker, such as the main source node, is tripped by the relay protection device to cut out the fault current. The clear control is also the start condition of centralized intelligence based fault location, isolation, and service restoration.

6.2.1.2 Fault Location, Isolation, and Service Restoration

Fault Location, Isolation, and Service Restoration may be carried out by local, distributed, or centralized intelligence.

- Relay protection.
  For rural areas, both the three-section type of over-current protection and over-current protections with time delay coordination are feasible. But for urban areas, only over-current protection with time delay coordination of no more than three stages is feasible. The control of fault location, isolation, and service restoration can be able to be

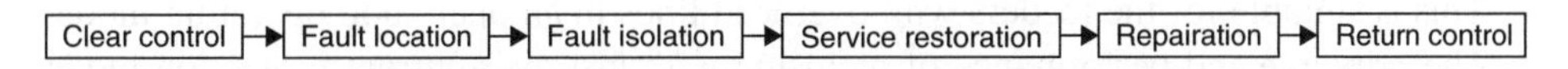

Figure 6.1 Six steps of the fault processing procedure

completed in no more than 1 sec. The fault region may not be located, however, and isolated to the smallest scope, and the restorable regions may not be fully restored. Besides, in some cases, coordination of over-current protection is not effective, for example, the sectionalizing switches on a trunk with very short lengths are difficult to coordinate.

- Automatic reclosing control.
  Automatic reclosing control is suitable for the feeders with overhead lines. Although automatic reclosing control cannot locate and isolate the fault, it may restore the service immediately in case of temporary faults. If the automatic reclosing control is only implanted on output circuit breakers in the substation, once a fault causes multistage tripping, only part of the feeder may be restored even if the reclosing is successful. But if the automatic reclosing controls are implanted on all of the action nodes with over-current protection, the loading of the feeder may all be restored even if a temporary fault causes multistage tripping.
- Backup automatic switching control.
  Although backup automatic switching control cannot locate and isolate the fault, once a fault occurs on the normal power supplying route, it may switch the load to the backup power supplying route within several seconds.
- FA based on recloser and voltage-delay type sectionalizers.
  This is suitable for the feeders with overhead lines in the open loop operating mode. The trip of a main source node, that is, the recloser, is needed to clear the fault current that caused the outage of the whole feeder and all sectionalizers on the feeder to trip due to loss of voltage. After a while, the main source node recloses for the first time, the sectionalizers close in sequence after being energized. For a temporary fault, the feeder will be restored within 1–2 min. For a permanent fault, the first reclosing is doomed to fail but the fault region is discovered. Then, the upstream of the feeder is restored by the reclosing of the second time, the downstream of the feeder is restored by the automatic closing of the loop switch, which causes the corresponding sectionalizers to close in sequence after being energized.

  The advantages of FA based on recloser and voltage-delay type sectionalizers lies in its simplicity, reliability, and no need for communication. It can locate and isolate the fault in the area surrounded with the sectionalizers. Both upstream and downstream of the fault may be restored. But it has the shortcomings of reclosing twice and the undesired close looping may be caused once the PT of the sectionalizer implanted on the loop switch is broken. Thus, sometimes the automatic function of loop switch is locked. Besides, the restoration schemes are unchangeable even in the cases of with the risk of causing over-current and therefore the outage area may be enlarged occasionally, especially in the area with loads increasing so fast.
- Reclosing with fast over-current protection mode.
  This is suitable for the feeders with overhead lines in an open loop operating mode. The principle and performance of reclosing with fast over-current protection mode are both similar to FA based on recloser and voltage-delay type sectionalizers, but

reclosing is only needed once; that is, the outage of the whole feeder only occurs once. Besides, all the sectionalizing switches should be circuit breakers while the sectionalizing switches of FA based on recloser and voltage-delay type sectionalizers are load switches.

- Fast healing approach based on neighbor communication.
  This is suitable for both overhead lines and cables in both the open loop operating mode and closed loop operating mode. The fault area can be located and isolated immediately and the healthy areas have almost no influence of the fault. But high speed communication and reliability are needed. Also, the sectionalizing switches should be circuit breakers.
- DAS based on centralized intelligence.
  The minimum area of fault location is the monitoring region surrounded by monitoring nodes, such as fault indicators. The minimum area of fault isolation is the action region surrounded by action nodes, which are usually the switches with FTUs. Generally, the fault location area of DAS based on centralized intelligence can be much smaller than that of other fault processing approaches. The fault isolation of DAS based on centralized intelligence can be satisfied even in cases where other fault processing approaches are not effective, such as a fault occurs on a trunk cable with numerous sections of short length divided by sectional switches such that none of the approaches of over-current protection, FA based on recloser and voltage-delay type sectionalizers, or reclosing with fast over-current protection mode, are feasible.

  Since global information can be collected, DAS based on centralized intelligence may optimize the service restoration schemes and generate correction schemes when some remote controls fail. Also, the DAS may deal with any faults, including cases of modeled fault processing, large area breakdown due to bus voltage losses, simultaneous multiple faults, the fault on a distribution grid with large-scale DGs, and so on. The remote control of a lot of switches may be arranged in methodical steps to guarantee safety.

  Centralized intelligence based DAS fault processing needs a rather long time period, typically, several minutes. Also, communication systems are needed to cover the DAS master stations and the FTUs, which usually requires a large sum of money.

6.2.1.2.1 Return Control

The return control to normal operation mode after repair must be carried out by the centralized intelligence based DAS. Since the global information can be collected, the DAS may optimize the return control schemes and arrange the operation steps to guarantee safety. When some remote controls fail, it can generate correction schemes as well.

Of course, in case of a temporary fault, the fault processing approaches of automatic reclosing control, FA based on recloser and voltage-delay type sectionalizers, reclosing with the fast over-current protection mode and the fast healing approach based on neighbor communication may also return the distribution grid to its normal operation mode. But in the case of permanent faults, they cannot do so.

A comparison of various fault processing approaches is shown in Table 6.1.

6.2.1.3 Basic Principles of Coordination of Automatic Fault Processing Approaches

According to Table 6.1, we may draw the following conclusions about automatic fault processing:

- The unique measure to cut out the fault current immediately – although the fast healing approach based on neighbor communication has the function of clear control, it is also based on relay protection, which is an essential methodology for automatic fault processing in distribution grids, but the fault location and isolation are somewhat rough.
- The unique measure to return to the normal operation mode – although FA based on recloser and voltage-delay type sectionalizers, reclosing with fast over-current protection mode, and the fast healing approach based on neighbor communication retain part of the function of return control, they are not effective in case of permanent faults, DAS based on centralized intelligence is an essential methodology for automatic fault processing of in grids.
- The methodologies based on local intelligence are faster and more reliable. Those based on distribution intelligence are simple and do not need of communication, except for the fast healing neighbor communication approach, which is fast and has little influence on healthy regions. Centralized intelligence based DAS can process complicated fault cases with more accurate fault location and optimized control steps and is very adaptable.

The basic principles of coordinating various fault processing methodologies are as follows:

- Once a fault occurs, local intelligence methodologies work first to clear the fault current immediately and roughly. As for the permanent faults, the fault is roughly located and isolated immediately by relay protections. As for the temporary faults, the automatic reclosing control restore all of the service immediately. In case the loads are with more than one power supplying route, the backup switching devices are helpful to make the restoration in seconds.
- A short time later, all of the fault information has been collected by the master of DAS, based on which the fault may be accurately located and the optimized control steps may be carried out to improve the service restoration result.
- The distributed intelligence without communication based methodologies can be used as the complement to DAS based on centralized intelligence to be installed in areas where the cost of communication is too high, or be installed on the branches.

Table 6.1 Comparisons of various fault processing approaches

Approaches		Suitable scopes	Clear control	Fault location	Fault isolation	Service restoration	Return control
Local intelligence	Relay protection	Cable/ Overhead	Fast	Fast, roughly	Fast, roughly	Fast, roughly	Can't
	Automatic reclosing	Overhead	Can't	Can't	Can't	Fast for only temporary fault	Fast for only temporary fault
	Backup switching	Cable/ Overhead	Can't	Can't	Can't	Fast for loads with multi-power-supplying routes	Most can't
Distributed intelligence	Recloser and voltage-delay sec-tionalizers	Overhead Open loop	Can't	Can locate the fault Temporary outage Strike twice Longer processing time Without communication	Can isolate the fault Temporary outage Strike twice Longer processing time Without communication	Can restore service Poor adaptability Risk of close loop Without communication	Can return to normal operation mode for only temporary fault Without communication
	Reclosing with fast protection	Overhead Open loop	Can't	Can locate the fault Temporary outage Strike once Shorter processing time Without communication	Can isolate the fault Temporary outage Strike once Shorter processing time Without communication	Can restore service Poor adaptability Risk of close loop Without communication	Can return to normal operation mode for only temporary fault Without communication

(*continued*)

Table 6.1 (*Continued*)

Approaches		Suitable scopes	Clear control	Fault location	Fault isolation	Service restoration	Return control
Centralized intelligence	Fast healing	Cable/Overhead Open/Closed loop	Fast	Fast No temporary outage No more strike Need communication	Fast No temporary outage No more strike Need communication	Can restore service Poor adaptability Risk of close loop Need communication	Can return to normal operation mode for only temporary fault Need communication
	DAS	Cable/Overhead Open/Closed loop	Can't	Locate the fault as accurate as the monitoring regions Temporary outage No more strike Longest processing time Need communication	Isolate the fault as accurate as the monitoring regions Temporary outage No more strike Longest processing time Need communication	Can restore service Good adaptability Optimized scheme and operation steps Need communication	Can return to normal Good adaptability Optimized scheme and operation steps Need communication

- The fast healing approach based on neighbor communication is only suitable for installation on feeders with a high reliability requirement.

With the coordination of various fault processing methodologies, not only does the performance of fault processing remarkably improve, the robustness is also greatly increased. For example, in case of the override trip or multistage tripping due to coordination failure of over-current protection, DAS based on centralized intelligence may locate the fault, isolate the fault region, and restore the service to correct the override trip or multistage tripping. On the other hand, in case where centralized intelligence based DAS fails due to a communication barrier, the fault can be roughly located and isolated by relay protection, and sometimes part of or even the whole healthy region may be restored, especially when the fault is on a lateral or branch.

An example of coordination of DAS and relay protections for fault processing is shown in Fig. 6.2. The distribution grid is based on cables where all switches are action nodes with FTUs. The squares indicate the circuit breakers, the circles indicate the load switches, the solid ones are in the closed states, while the hollow ones are open.

The circuit breakers in the substations, that is, S_1 and S_2, are installed with the time-delay instantaneous current protection and over-current protection with set value of time delay of 0.3 and 0.5 sec., respectively. The circuit breakers indicated by big squares are installed with instantaneous current protection without delay and over-current protection with set-values of time delay of 0.2 sec. The circuit breakers indicated by small squares are installed with instantaneous current protection without delay and over-current protection with set-values of time-delay of 0 sec. Thus, coordination of over-current protection can be as much as three stages while the coordination of instantaneous current protection can be two stages. Also, since the set-values of time-delay of the instantaneous current protection C_2 and C_1 are both 0 s while their set-values of current are almost the same, override trip or multistage tripping of C_2 and C_1 is inevitable.

In the case where a two-phase short circuit fault occurs in region with C_1 as its enter point shown in Fig. 6.2(b), since the short-circuit current is smaller than the set-value of the instantaneous current protection of S_1, C_1 and C_2, only the over-current protections of S_1, C_2, and C_1 are started. The set-value of time-delay of C_1 is the shortest, it is tripped to clear the fault current and isolate the fault region while there is almost no influence on the other regions of the feeder, which is shown in Fig. 6.2(b).

In the case where a three-phase short circuit fault occurs in the region with C_1 as its enter point shown in Fig. 6.2(c), since the short-circuit current is very large, the instantaneous current protections and over-current protections of S_1, C_2, and C_1 are all started. As the set-values of time-delay of the instantaneous current protections of C_2 and C_1 are both 0 sec., C_2 and C_1 are all tripped almost simultaneously causing a multistage tripping, the fault current is cleared and the fault is isolated but not within the smallest area. There is almost no influence on the other regions of the feeder, which is shown in Fig. 6.2(c). After a while, the exact fault region is located by the centralized intelligence based DAS. Then C_2 is closed by the

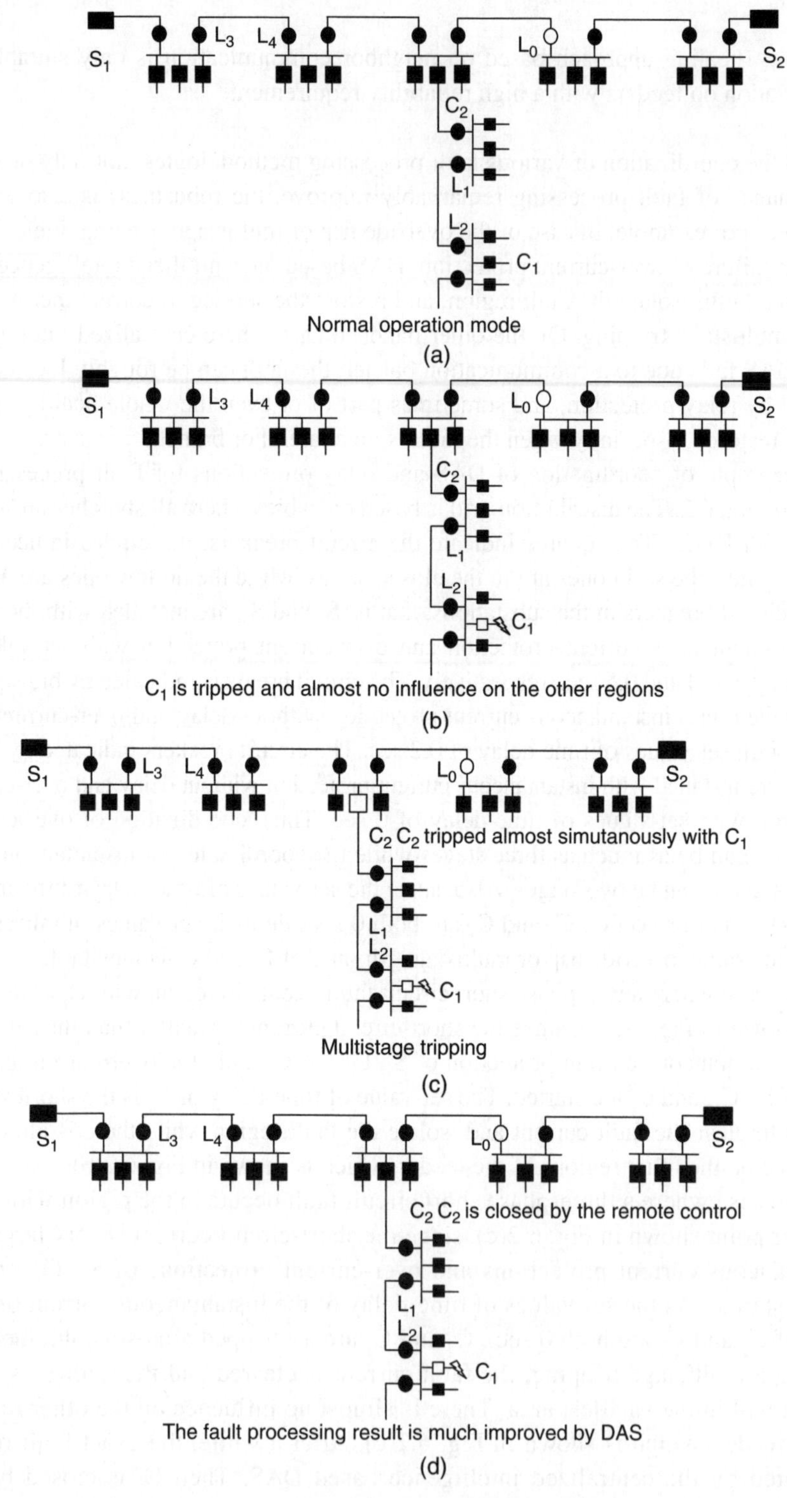

Figure 6.2 An example of coordination of DAS and relay protection for fault processing

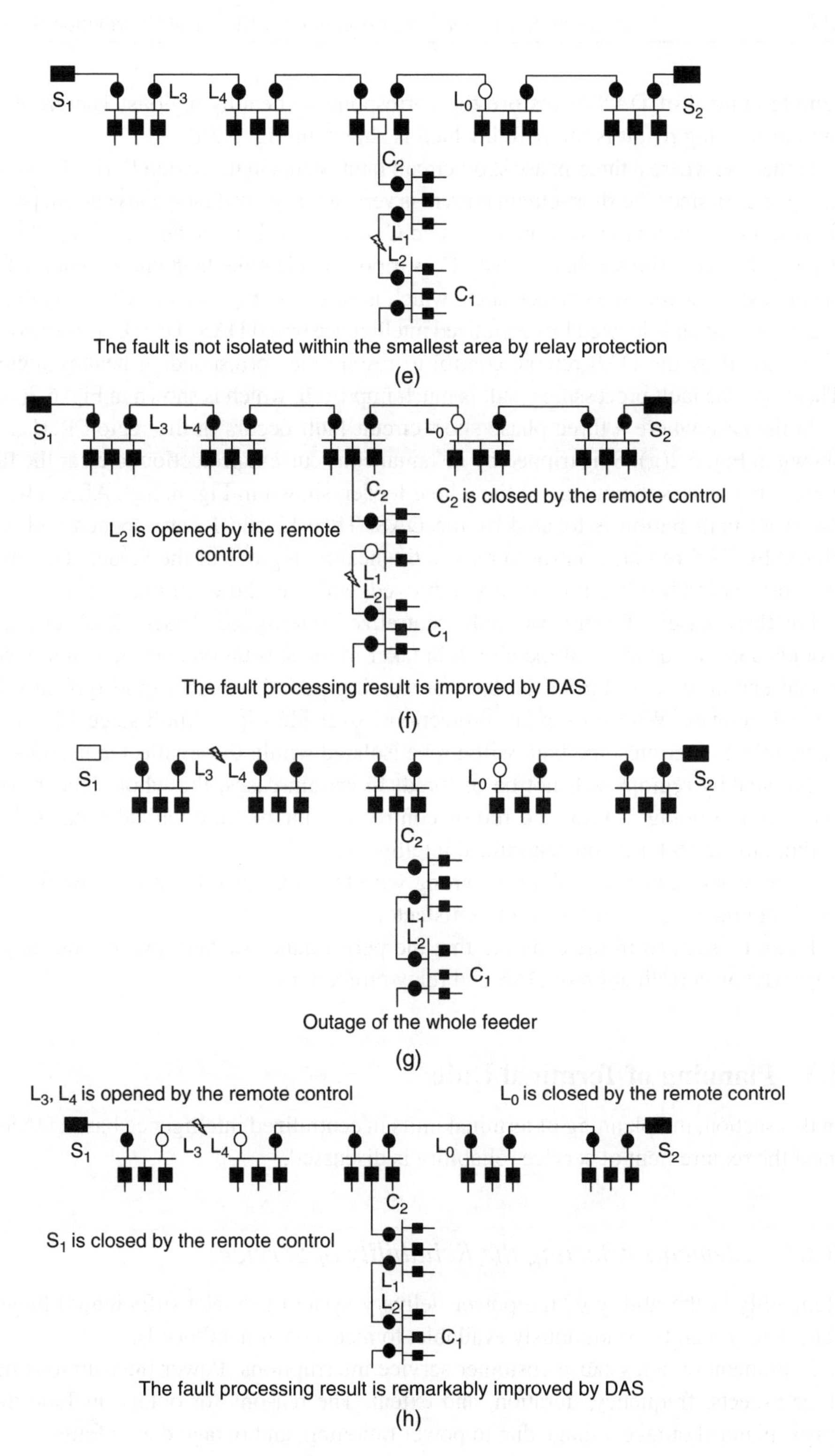

Figure 6.2 (*continued*)

remote control of DAS to restore the corresponding healthy regions. Therefore, the fault processing result is improved, which is shown in Fig. 6.2(d).

In the case where a three-phase short circuit fault occurs in the region Re(L_1, L_2) shown in Fig. 6.2(e), since the short-circuit current is very large, the instantaneous current protections and over-current protections of S_1 and C_2 both start. Due to the set-values of time-delay of C_2 being shorter than S_1, only C_2 is tripped to clear the fault current but the fault is not isolated within the smallest area, which is shown in Fig. 6.2(e). After a while, the exact fault region is located by centralized intelligence based DAS. Then L_1 is opened and C_2 is closed by the DAS remote control to restore the corresponding healthy regions. Therefore, the fault processing result is much improved, which is shown in Fig. 6.2(f).

In the case where a three-phase short circuit fault occurs in the region Re(L_3, L_4) shown in Fig. 6.2(g), S_1 is tripped by instantaneous current protection to clear the fault current but causes the outage of the whole feeder, shown in Fig. 6.2(g). After a while, the exact fault region is located by the DAS. Then L_3 and L_4 are opened and S_1 is closed by DAS remote control to restore the healthy regions of the feeder. Therefore, fault processing result is remarkably improved, which is shown in Fig. 6.2(h).

For these cases, if there was only centralized intelligence based DAS, any fault would cause the tripping of the circuit breaker in the substation and the whole feeder would endure outage during the fault processing period, which would typically last several minutes. With only relay protections, override trip or multistage tripping is inevitable. Sometimes, the fault will not be isolated within the smallest area and some of the healthy regions will not be restored. In certain cases, the whole feeder could even endure outage. Also, the return control cannot be automatically carried out without the DAS based on centralized intelligence.

To be worse, once something is wrong with the DAS or relay protection devices, the fault processing result would unsatisfactory.

It can be seen from the example that the performance of fault processing may be improved by coordination of DAS and relay protections.

6.3 Planning of Terminal Units

In this section, the planning of terminal units in centralized intelligence based DASs to meet the requirement of service reliability is discussed.

6.3.1 Elements Affecting the Reliability of Service

Reliability is the ability of the power delivery system to make sufficient voltage of satisfactory quality continuously available, to meet consumers' needs.

Equipment outages cause customer service interruptions. Power interruptions have three aspects: frequency, duration, and extent. The reasons for outage include three types: planned outage, outage due to power rationing, and outage due to faults.

Table 6.2 Fault ratios per kilometer of overhead lines and cables in China

	Overhead lines (times/100 km/year)	Cables (times/100 km/year)
2002	9.674	4.447
2003	8.343	4.059
2004	9.408	4.148
2005	9.62	4.27
2006	11.656	4.115
2007	9.404	3.553
2008	10.58	3.01
2009	9.19	3.55
Average	9.73	3.89

Outage due to power rationing should be reduced by expanding planning and engineering of the electric power grid, usually carried out satisfactorily. Thus, the ratio of outage due to power rationing in most areas is small, for example, less than 5% in China.

The ratios of planned outage are quite different between utilities. These are determined by the level of condition based maintenance and non-blackout working technology, the skills and arrangement of repair, extent of traffic jam, distance, surroundings of the feeder, and so on. The ratios of planned outage of utilities vary from 30–70% in China.

With the development of distribution grids and the maintenance and non-blackout working technology, outage due to faults increases to the most important reason for outage.

The fault processing technologies described in this book are of great help to reduce the area and time of the outage due to faults.

To simplify the problem, the per-unit fault ratio is usually indicated by the times of fault on unit length of the feeder, which can be obtained by statistics. For instance, the fault ratios per kilometer of overhead lines and cables in China are shown in Table 6.2, which are obtained by statistical analysis of reliability data from 2002 to 2009.

The other important index is the averaged outage duration time of one fault, which consists of the time to find out the exact position of the fault and the time of repair. For instance, the averaged outage duration times due to faults in China are shown in Table 6.3, which are obtained by statistical analysis of reliability data from 2005 to 2009.

6.3.2 Cost-Benefit Analysis of Action Node Planning

For a feeder with n customers with the total load of P, it is connected to other feeders and the $N - 1$ criterion is satisfied. We suppose that sectionalizing switches are sufficient, the fault ratio of the feeder is F, the averaged outage duration time of one fault is T, the cost of a FTU is C, and the number of FTUs to be installed is k, which divide the feeder into $k + 1$ action regions.

Table 6.3 The averaged outage duration time of one fault in China

	Averaged outage duration time of one fault (hour/times)
2005	3.30
2006	2.63
2007	2.40
2008	3.26
2009	2.28
Average	2.77

6.3.2.1 The Case of the Amount of Customers Evenly Distributed

Assuming that the amount of customers is evenly distributed, the number of customers in each action region is $n/(k+1)$.

Since the $N-1$ criterion is satisfied, the benefit of installing k FTUs is to maintain the service of the action regions except for the fault action region before repairation being completed, which can be formulated as Equation (6.1).

$$B_1(k)=\frac{nkFT}{k+1}(\text{hour}\cdot\text{customer}) \tag{6.1}$$

It can be seen from (6.1) that the increase in benefit becomes weaker and weaker with the increase in the amount of FTUs installed.

The cost of installing k FTUs is approximately

$$C(k)=kC(\$) \tag{6.2}$$

The cost-benefit ratio of installing k FTUs is

$$BC_1(k)=\frac{nFT}{(k+1)C}(\text{hour}\cdot\text{customer}/\$) \tag{6.3}$$

It can be seen from (6.3) that $BC_1(k)$ is a monotonically decreasing function, that is, the more FTUs installed, the lower the cost-benefit ratio is.

6.3.2.2 The Case of Evenly Distributed Loads

Assuming that the loads are evenly distributed, the load in each action region is $P/(k+1)$.

The benefit of installing k FTUs is

$$B_2(k)=\frac{PkFT\lambda}{k+1}(\$) \tag{6.4}$$

Where, λ is the benefit of unit load in 1 h.

It can be seen from (6.4) that the increasing of the benefit becomes weaker and weaker with the increasing amount of FTUs installed.

The cost-benefit ratio of installing k FTUs is

$$BC_2(k) = \frac{PFT\lambda}{(k+1)C} \tag{6.5}$$

It can be seen from (6.5) that $BC_2(k)$ is also a monotonically decreasing function, that is, the more FTUs installed, the lower the cost-benefit ratio is.

The net benefit of installing k FTUs is

$$B_3(k) = \frac{PkFT\lambda}{k+1} - kC(\$) \tag{6.6}$$

The net cost-benefit ratio of installing k FTUs is

$$BC_3(k) = \frac{PFT\lambda}{(k+1)C} - 1 \tag{6.7}$$

It can be seen from (6.7) that $BC_3(k)$ is still a monotonically decreasing function, that is, the more FTUs installed, the lower the cost-benefit ratio is.

In conclusion, from the viewpoint of cost-benefit ratio, the scheme of installing only one FTU has the highest cost-benefit ratio. But the scheme of installing FTUs should also meet the requirement of service reliability. Thus, the number of FTUs installed is the minimum amount to meet the requirement of service reliability, which may have the highest possible cost-benefit ratio.

6.3.3 Planning the Amount of Terminal Units to Meet the Requirement of Service Reliability

Terminal units can be classified into two types; the FTUs with the function of tele-control, tele-metering, and tele-indication, and the fault indicators with the function of tele-indication and inaccurate tele-metering.

To simplify the problem, we suppose that each terminal unit controls or monitors only one node. In fact, although the pole mounted FTU on overhead lines may control only one node and the fault indicator may monitor only one node, the FTU installed in a ring main unit cabinet sometimes may control more than one node, which will be discussed later.

Assuming that the amount of customers is evenly distributed and the number of terminal units to be installed is k, which divides the feeder into $k + 1$ regions, the number of customers in each region is $n/(k + 1)$.

The duration time of fault processing consists of three parts shown in Equation (6.8)

$$T = t_1 + t_2 + t_3 \tag{6.8}$$

Where, t_1 is the time of searching the fault position, t_2 is the time to isolate the fault region and restore the service of the healthy regions by manual work, and t_3 is the repair time of the failed devices corresponding to the fault.

6.3.3.1 Only-FTU Mode

For the only-FTU mode, all the nodes needed to equip terminal units are installed with FTUs with the function of tele-control, tele-metering, and tele-indication. Also, each node with FTU should be a switch with an electric operating mechanism. Thus, the cost of the only-FTU mode is much higher but all the operations in fault processing are automatic by remote control. Thus, service restoration of the healthy regions is fast.

For the only-FTU mode, both t_1 and t_2 are approximately zero, then we have

$$T = t_3 \tag{6.9}$$

In case of the $N-1$ criterion being satisfied and with k FTUs installed, the Average Service Availability Index ($ASAI_3$) to only include the outage due to a fault can be deduced as Equation (6.10) according to the definition by the ASAI

$$ASAI_3 = \frac{8760n - \sum_{i=1}^{k+1} \frac{nt_3 f_i}{k+1}}{8760n} = 1 - \frac{\sum_{i=1}^{k+1} t_3 f_i}{8760(k+1)} \tag{6.10}$$

Where, f_i is the fault ratio of the *i-th* region.

Assuming that the total fault ratio of the feeder is F and each region has the same fault ratio of f, we have

$$f_i \approx f = \frac{F}{k+1} \quad (0 < i \le k+1) \tag{6.11}$$

Equation (6.10) can be re-written as

$$ASAI_3 = 1 - \frac{\sum_{i=1}^{k+1} t_3 F/(k+1)}{8760(k+1)} = 1 - \frac{t_3 F}{8760(k+1)} \tag{6.12}$$

If the desired ASAI to only include the outage due to fault is no less than A, that is, $ASAI_3 \geq A$, we may obtain the number of FTUs needed to be installed as

$$k \ge \frac{t_3 F}{8760(1-A)} - 1 \quad (k \ge 0) \tag{6.13}$$

Besides, all of the loop switches should also be installed with FTUs, we assume that the amount of the FTUs installed on loop switches is k_0

It can be seen from Equation (6.13) that the number of FTU needed to be installed depends on the desired ASAI to only include the outage due to fault, the repair time of the failed devices corresponding to the fault, and the fault ratio.

6.3.3.2 Only Fault Indicator Mode

For the only fault indicator mode, all of the nodes needed to equip terminal units are installed with fault indicators with the function of tele-indication and inaccurate tele-metering. A fault indicator may be installed to monitor a switch with or without an electric operating mechanism; it also may be installed on a certain position of a branch or lateral without any switch. Thus, the cost of the only fault indicator mode is much lower. But all the operations in fault processing must be done by manual labor, causing the service restoration of the healthy regions to be slow. The fault location process is as fast as that of the only-FTU mode.

For the only fault indicator mode, t_1 is approximately zero, then we have

$$T = t_2 + t_3 \tag{6.14}$$

In case of the $N-1$ criterion being satisfied and with k fault indicators installed, the Average Service Availability Index ($ASAI_2$) to only include the outage due to fault can be deduced as Equation (6.15) according to the definition from the ASAI

$$ASAI_2 = \frac{8760n - nFt_2 - \sum_{i=1}^{k+1} \frac{nt_3 f_i}{k+1}}{8760n} = 1 - \frac{(k+1)Ft_2 + \sum_{i=1}^{k+1} t_3 f_i}{8760(k+1)} \tag{6.15}$$

In Equation (6.15), nFt_2 is the loss of outage (hour·customer) due to service interruption of the whole feeder, which lasts the time t_2 before the fault is isolated, because fault indicators lack the tele-control function.

Assuming that the total fault ratio of the feeder is F and each region has the same fault ratio of f, we have

$$ASAI_2 = 1 - \frac{F}{8760}\left(t_2 + \frac{t_3}{k+1}\right) \tag{6.16}$$

If the desired ASAI to only include the outage due to fault is no less than A, that is, $ASAI_2 \geq A$, we have

$$1 - A \geq \frac{F}{8760}\left(t_2 + \frac{t_3}{k+1}\right) \tag{6.17}$$

From Equation (6.17), we may obtain the number of fault indicators needed be to installed as

$$k \geq \frac{t_3 F}{8760(1-A) - t_2 F} - 1(k \geq 0) \tag{6.18}$$

It can be seen from Equation (6.18) that the number of fault indicators needed depends on the desired ASAI to only include the outage due to fault, the time to isolate the fault region, and restore the service of the healthy regions manually, the repair time for the failed devices corresponds to the fault and the fault ratio.

It is worth noting that Equation (6.18) is deduced in the condition of $k+1>0$, thus, once the value of $\frac{t_3 F}{8760(1-A) - t_2 F} - 1$ is less than -1, it means that no feasible solution exists.

6.3.3.3 Relay Protection with Terminal Units Mode

For relay protection with terminal units mode, relay protection devices are installed on the branches or laterals with higher fault ratios and coordinated with relay protection devices in the substations to reduce the influence of outage due to faults, which is equivalent to decreasing the fault ratio of the feeder. FTUs or fault indicators are also installed to further improve the reliability.

Assuming that the ratio of amount of branch or lateral faults to the fault of the whole feeder is Ψ, the ratio of relay protection devices installed on branches and laterals is μ, the fault ratio of the whole feeder may decrease to F' as shown in Equation (6.19)

$$F' = F - \psi F \mu \tag{6.19}$$

Replace the F in Equation (6.13) with F' and we may obtain the number of FTUs needed to be installed as shown in Equation (6.20)

$$k \geq \frac{t_3 F'}{8760(1-A)} - 1(k \geq 0) \tag{6.20}$$

Also, all the loop switches should also be installed with FTUs.

Replace the F in Equation (6.18) with F' and we may obtain the number of fault indicators needed to be installed as shown in Equation (6.21)

$$k \geq \frac{t_3 F'}{8760(1-A) - t_2 F'} - 1(k \geq 0) \tag{6.21}$$

6.3.3.4 Hybrid Mode of FTUs and Fault Indicators

As for the hybrid mode of FTUs and fault indicators, some of the terminal units are FTUs, the other terminal units are fault indicators. The cost of the hybrid mode of FTUs and fault indicators is moderate, where the area of fault isolation may be larger than the located fault region.

Assuming that the summation of FTUs and fault indicators on a feeder is k and the fault indicators are evenly inserted in the area surrounded by FTUs, source nodes, and ending nodes with h fault indicators in each, we have

$$k = (k_1 + 1)h + k_1 \tag{6.22}$$

wher, k_1 is the number of FTUs.

Comparing Equation (6.16) with (6.12), we may find that they have a common part of $\frac{t_3 F}{8760(k+1)}$. In another word, the sum of all of the influence on service reliability of FTUs and partial influence on service reliability of fault indicators is $\frac{t_3 F}{8760(k+1)}$.

The other influence on service reliability of fault indicators in Equation (6.16) is $\frac{Ft_2}{8760(k_1+1)}$. It is due to the service interruption of the whole feeder, which lasts the time of t_2 before the fault is isolated, because fault indicators lack the tele-control function.

Therefore, as for hybrid mode of FTUs and fault indicators, the Average Service Availability Index ($ASAI_2$) to only include the outage due to fault can be deduced as Equation (6.23)

$$ASAI_{3,2} = 1 - \frac{F}{8760}\left(\frac{t_2}{k_1+1} + \frac{t_3}{k+1}\right) = 1 - \frac{F}{8760}\left(\frac{t_2}{k_1+1} + \frac{t_3}{(1+h)(k_1+1)}\right) \geq A \tag{6.23}$$

Thus, we may obtain the number of FTUs needed as

$$k_1 \geq \frac{F\left[(1+h)t_2 + t_3\right]}{8760(1-A)(1+h)} - 1 \left(k_1 \geq 0\right) \tag{6.24}$$

Similar to Equation (6.18), once k_1 is less than zero, it means that no feasible solution exists.

We may also obtain the number of fault indicators needed as

$$k_2 = (k_1 + 1)h \tag{6.25}$$

Also, all of the loop switches should also be installed with FTUs, we assume that the amount of the FTUs installed on loop switches is k_0.

As for hybrid mode of FTUs and fault indicators, the amounts of terminal units are different with the varying value of h. The costs of these feasible schemes are different. We should choose the scheme with the lowest cost.

6.3.3.5 Discussions

In previous paragraphs, the planning of amount of terminals in typical cases is investigated. In this paragraph, some problems are discussed.

6.3.3.5.1 Hybrid Feeder of Cables and Overhead Lines

In previous paragraphs, the fault ratio per kilometer is assumed to be the same, in these cases the feeder is constituted by merely cables or overhead lines.

For the hybrid feeder of cables and overhead lines, if it mainly constitutes cables with a few overhead lines, it can be approximately regarded as a cable based feeder. Conversely, if the feeder mainly constitutes overhead lines with a few cables, it can be approximately regarded as overhead line based feeder. If both the cable part and the overhead line part cannot be ignored, they should be planned according to the methods described in previous paragraphs, respectively.

6.3.3.5.2 Radial Feeders

In previous paragraphs, the planning method in case of satisfaction of the $N-1$ criterion is described, which is not applicable for radial feeders.

Since radial feeders are mainly in rural areas, fault indicators are the first choice instead of FTUs. Assuming that k fault indicators are installed on the trunk dividing the feeder into $k+1$ sections and the desired ASAI to only include the outage due to fault is no less than A, we have

$$ASAI_2 = 1 - \frac{(k+2)t_3F + 2(k+1)t_2F}{2\times 8760(k+1)} \geq A \tag{6.26}$$

From Equation (6.26), we may obtain the number of fault indicators needed as

$$k \geq \frac{t_3F}{17520(1-A) - t_3F - 2t_2F} - 1 \left(k \geq 0\right) \tag{6.27}$$

If the relay protection of $k+1$ stages and automatic reclosing control equipped on the trunk, which is often feasible for the feeder in rural areas, the time of t_2 may be reduced in fault processing. Thus, we have

$$k \geq \frac{t_3F}{17520(1-A) - t_3F} - 1 \left(k \geq 0\right) \tag{6.28}$$

As for Equation (6.27) and (6.28), once k is less than zero, it means that no feasible solution exists.

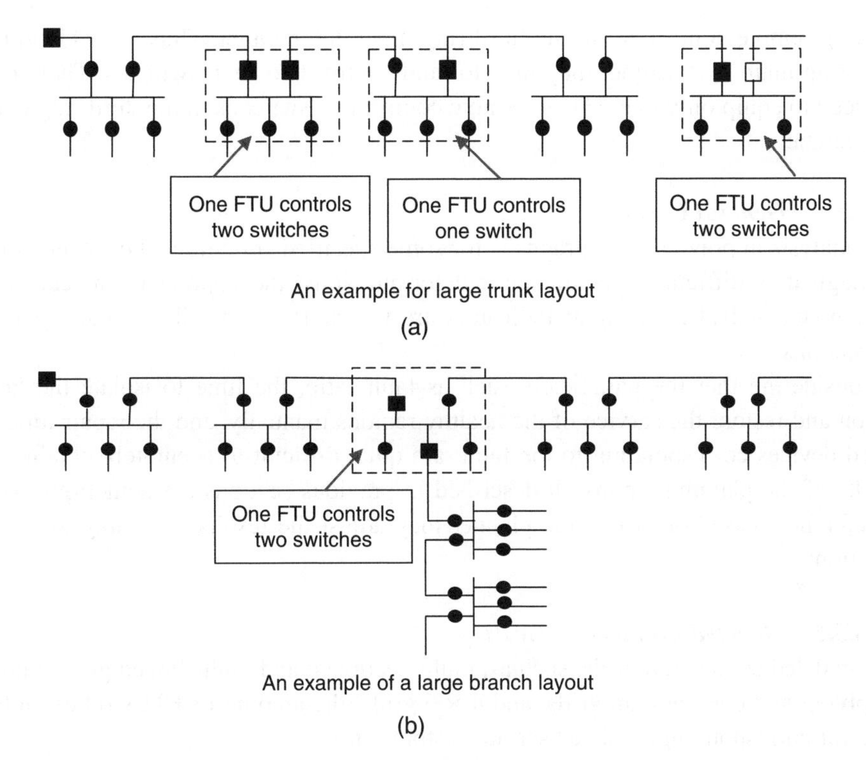

Figure 6.3 Examples to explain the amount of FTUs for cable feeders

6.3.3.5.3 FTU in a Ring Main Unit Cabinet

A FTU installed in a ring main unit cabinet sometimes may control more than one node, because the FTU is often equipped to monitor and control many switches.

Examples are shown in Fig. 6.3(a) and (b), respectively, in which the squares indicate the switches that have remote control ability, while the circles indicate the switches without the function of remote control. The solid ones indicate the corresponding switches are closed while the hollow ones indicate the corresponding switches are open.

A large trunk layout is shown in Fig. 6.3(a), in which the feeder is divided into five sections by four switches with remote control function. Two switches needing remote control are in the second ring main unit cabinet. Thus the FTU in the second ring main unit cabinet may monitor and control two such switches. Since only one switch needing remote control is in the third ring main unit cabinet, the FTU in the third ring main unit cabinet may monitor and control only one switch. The fourth switch needing remote control and the loop switch that also needs remote control are in the fifth ring main unit cabinet. Thus the FTU in the fifth ring main unit cabinet may monitor and control two such switches. Therefore, we need to equip three FTUs to remote control five switches distributed in three ring main unit cabinets.

A large branch layout is shown in Fig. 6.3(b), in which the feeder is divided into three sections by two switches with the function of remote control. These two switches needing remote control are in the third ring main unit cabinet. Thus the FTU in the third ring main unit cabinet may monitor and control two such switches. Therefore, we need to equip only one FTU to remote control two switches in the third ring main unit cabinet.

6.3.3.5.4 General Cases

The analysis in previous paragraphs is for some specified conditions. For other cases, although it is difficult to obtain unified formulations, the approaches to calculate the amount of FTUs and fault indicators are the same as describe in the previous subsections.

Considering that the parameters such as fault ratio, the time to isolate the fault region and restore the service of the healthy regions manually, and the repair time of failed devices corresponding to the fault, are quite difficult to accurately obtain, the results of the planning approach described in previous sections are sometimes used as reference, based on which the planner may adjust the results according to actual situations.

6.3.3.5.5 Modeled Connection Grids

For modeled connection grids, such as, multi-sectioned and multi-linked grids, multi-supplying and one back-up grids, and 4 × 6 grids, the amount of FTUs relies on the need for corresponding modeled service restoration.

6.3.3.6 An Example of Planning

A city can be divided into three types of area according to the desired reliability of service. Supposing that the desired Average Service Availability Indexes (ASAI) of A-type, B-type, and C-type areas are 99.99%, 99.965%, and 99.897%, respectively.

The city distribution grids consist of 590 feeders. The A-type area consists of 50 cables. The B-type area consists of 60 cables, 80 insulated overhead lines, and 120 hybrid feeders. The C-type area consists of 180 un-insulated overhead lines.

The load reach of the feeders in the A-type area and B-type area is about 5 km while that in the C-type area is about 10 km.

The fault ratios per kilometer of un-insulated overhead lines, cables, insulated overhead lines, and hybrid feeders are 0.1 (times/km.year), 0.04 (times/km.year), 0.07 (times/km.year), and 0.07 (times/km.year), respectively.

For the utility to supply the city distribution grid, the time to isolate the fault region and restore the service of the healthy regions by manual work is 1 h/time for the downtown area and 2 h/time for the suburb. The time of the repair of the failed devices corresponding to the fault is 4 h/time for the downtown area and 6 h/time for the suburb.

The ratios of outage due to fault of the utility are 80% for an A-type area, 60% for a B-type area, and 22% for C-type area. Therefore, the Average Service Availability

Indexes to only include the outage due to fault are 99.992% for an A-type area, 99.979% for a B-type area, and 99.977% for a C-type area.

These parameters may be obtained by statistical approaches.

It can be evaluated that if neither FTUs nor fault indicators are installed, the ASAI in the A-type area would be 99.9585%, the ASAI in the B-type area with cable feeders would be 99.9585%, while that with insulated overhead lines and hybrid feeders would be 99.9274%, the ASAI in the C-type area would be 99.7924%, which cannot reach the desired ASAI. Thus, FTUs or fault indicators should be installed in the A-type, B-type, or C-type area.

For the A-type area, the 50 cable based feeders are planned to form 25 couples of loop grids to meet the $N-1$ criterion requirement. The feeders are only FTUs installed. The 25 loop switches should equip FTUs in ring main unit cabinets. According to Equation (6.13), each feeder should equip one FTU in ring main unit cabinets. Thus, altogether, we need 75 FTUs in ring main unit cabinets for the A-type area. Also, no relay protection device should be implanted on the sectionalizing switches and loop switches of the feeder. All the sectionalizing switches and loop switches may be load switches instead of circuit breakers.

As for the 60 cable based feeders in the B-type area, they are planned to form 30 couples of loop grids to meet the $N-1$ criterion requirement. The feeders are only installed FTUs. The 30 loop switches should equip FTUs in ring main unit cabinets. According to Equation (6.13), each feeder should equip one FTU in ring main unit cabinet. Also, no relay protection device should be implanted on the sectionalizing switches and loop switches of the feeder. All of the sectionalizing switches and loop switches may be load switches instead of circuit breakers. Thus, altogether, we need 90 FTUs in ring main unit cabinets for B-type area cable based feeders.

As for the 120 hybrid feeders and 60 insulated overhead line based feeders in the B-type area, they are planned to form 90 couples of loop grids to meet the $N-1$ criterion requirement. The feeders are only installed FTUs. The 90 loop switches should equip FTUs. According to Equation (6.13), each feeder should equip one FTU. Also, no relay protection device should be implanted on the sectionalizing switches and loop switches of the feeder. All of the sectionalizing switches and loop switches may be load switches instead of circuit breakers. Thus, altogether, we need 270 FTUs for hybrid feeders and insulated over-head line based feeders of B-type area. Assuming that one-third of the FTUs are the type in the ring main unit cabinets, the others are pole mounted FTUs.

For the 180 un-insulated over-head line based feeders in C-type area, they are also planned to form 90 couples of loop grids to meet the $N-1$ criterion requirement. The feeders are only installed fault indicators. All of the branches and laterals with higher fault ratios are equipped with relay protection devices, automatic reclosing control, and circuit breakers. Assuming that three suits of such devices and fault indicators are equipped on each feeder, these reduce the fault ratio of each feeder by one-third.

Table 6.4 The results of planning

Terminal units and other devices	A	B	C	Total
FTUs in ring main unit cabinets	75	180	0	255
Pole mounted FTUs	0	180	0	180
Pole mounted fault indicators	0	0	720	720
Relay protection devices	0	0	540	540
Automatic reclosing controller	0	0	540	540
Fiberoptics (ONU)	75	360	0	435
GPRS devices	0	0	720	720

According to Equation (6.21), each feeder should equip one fault indicator on the trunk. Also, all of the sectionalizing switches and loop switches on the trunk may be load switches instead of circuit breakers. Thus, we need 180 fault indicators on the trunk for a C-type area, 540 suites of relay protection devices, automatic reclosing control, fault indicators, and circuit breakers on the branches and laterals. All of the terminal units are pole-mounted.

For FTUs, we chose fiberoptic communication media while for fault indicators, we chose GPRS.

The results of planning are shown in Table 6.4.

After engineering according to the planning results, the ASAI in the A-type area would be 99.9917%, in the B-type area with cable feeders 99.9689%, while that with insulated overhead lines and hybrid feeders would be 99.9728%, the ASAI in the C-type area would be 99.9222%, which can reach the desired ASAI. Thus, the planning is effective.

6.4 Verification of the Property of Fault Processing

The fault processing procedure needs the coordination of the corresponding devices. Thus, the performance of fault processing is rather difficult to verify. In the 1990s, centralized intelligence based distribution automation systems were constructed in almost every province of China, few of which worked well. One of the major reasons is the lack of testing methodology, causing fault processing properties that needed to be verified to have to wait a long time for a fault to occur on the feeder. Thus, the defects of the system cannot be found in time. Even if a certain fault is correctly processed, the fault processing procedure may also fail in the case where another fault occurs due to the diversity of various faults.

A series of testing methodologies and corresponding devices have been developed by the research team at the Shaanxi Electric Power Research Institute, which will be described in this section.

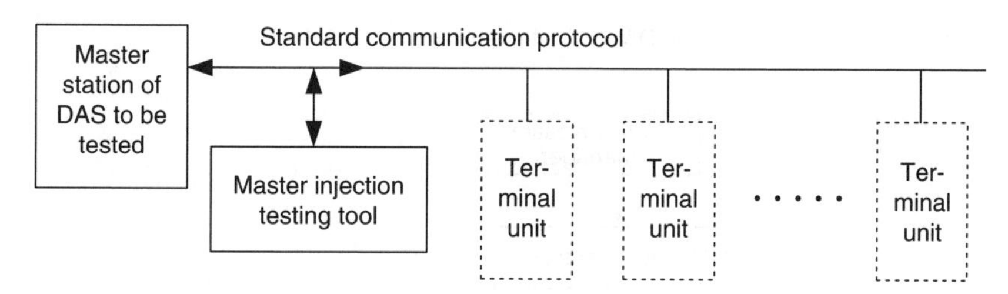

Figure 6.4 The master injection testing methodology

6.4.1 Master Injection Testing Methodology and the Testing Tool

6.4.1.1 Basic Principle

The *master injection testing methodology* is shown in Fig. 6.4.

This methodology aims to inject the information reflecting the fault phenomenon into the master station of a centralized intelligence based DAS to be tested by the master injection testing tool, which can set up various fault phenomena; such as different fault positions, load scenes, relay protection action, operation by mistake, and so on, and simulate large amount of FTUs and fault indicators to execute the information exchanging with the master station using standard protocols instead of actual terminal units to verify the performance of the fault processing procedure. In other words, the master injection testing tool provides a simulated fault environment for testing and responding to remote control orders from the master station of a DAS, thus, the fault processing procedure may be continuously carried out and the steps are recorded so as to evaluate the performance of the fault processing procedure.

6.4.1.2 Master Injection Testing Tool

The master injection testing tool consists of seven modules; a load simulator to calculate the load flow of a distribution grid, a fault simulator to produce the fault phenomenon, a modeling and configuration tool, a real time database, a protocol translator, a communication manager, and a human-computer interface, which is shown in Fig. 6.5.

6.4.1.2.1 Load Simulator

The load simulator has the functions of network topology analysis and load flow calculation, both of which are based on the simplified model of distribution grids described in Chapter 4.

- The network topology analysis sub-module establishes the network topology of the distribution grid according to the states of switches in the real-time database.

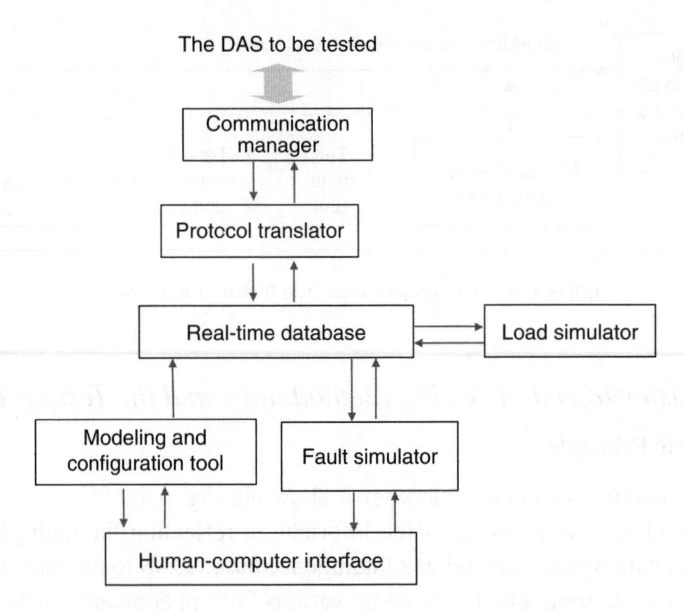

Figure 6.5 Structure of the master injection testing tool

- The load flow calculation sub-module calculates the load flow according to the distributed load information from the regions in the real-time database to obtain the loads flowing through each node to simulate real-time information from the action and monitoring nodes collected by terminal units and then saves them in the real-time database.

6.4.1.2.2 Real-Time Database

The real-time database has the following functions:

- Renewing the states of switches and the loads distributed from the regions according to the order from the protocol translator, the modeling and configuration tool, and the fault simulator.
- Renewing the loads flowing through each node.
- Renewing the fault information according to the orders from the fault simulator.

6.4.1.2.3 Modeling and Configuration Tool

The modeling and configuration tool has the following functions:

- Chart module integration based modeling of distribution grids adopting the simplified model of distribution grids described in Chapter 4.
- Configuration of switches, including their types, such as, circuit breaker, load switch, recloser, and so on, and states, such as closed or open.

- Configuration of load distributed from each region, which is described by load curves and with random fluctuation.
- Configuration of the terminal units and the corresponding configuration information of tele-control, tele-metering, and tele-indication.

6.4.1.2.4 Fault Simulator

The fault simulator has the following functions:

- Configuration of fault scenarios, such as fault positions (one or more), fault types (permanent or temporary), the switches refuse to act or not, reclosing is permitted or not, fault information missing or not, override trip or not, and so on.
- Produce the fault information and send it to the real-time database according to the configuration of the fault scenes and network topology.

The flowchart of fault simulation is shown in Fig. 6.6.

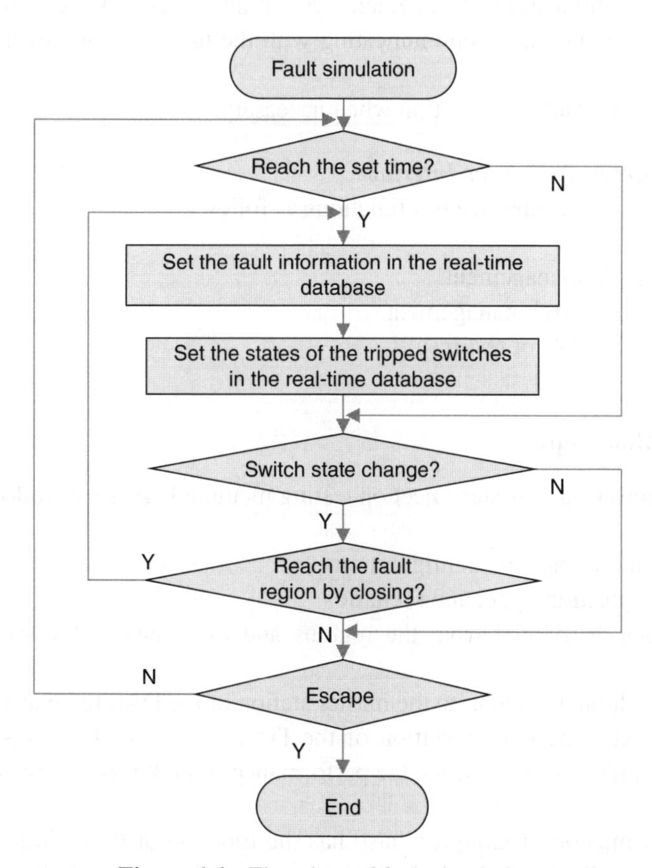

Figure 6.6 Flowchart of fault simulation

6.4.1.2.5 *Protocol Translator*

The protocol translator has functions as follows:

- Configuration of the protocols, such as DNP, IEC-61850, IEC60870–5-101, IEC60870–5-104, and so on.
- Generation of the upgoing message according to the information in the real-time database produced by the load simulator and the fault simulator.
- Translation of the downgoing message and putting the load and switch state information into the real-time database. When receiving a tele-control order, set or reset the state of the corresponding switch considering whether the switch set refuses to act.
- Keep the real-time data interaction between the real-time database and the master station of the DAS to be tested.

6.4.1.2.6 *Communication Manager*

The communication manager has functions as follows:

- Message communication management with multi-IP address to simulate a large amount of terminal units communicating with the master station of the DAS to be tested.
- Link monitoring and reconnection when necessary.

6.4.1.2.7 *Human-Computer Interface*

The human-computer interface has functions as follows:

- Input and output management.
- Operation and control management.
- Generation of a table of test results.

6.4.1.3 Testing Steps

The testing steps of the master injection testing methodology are as follows:

- Data input, modeling, and configuration.
- Set the fault position, type, and scenarios.
- Set the loads distributed from the regions and the states of the switches before the fault.
- Inject the fault phenomenon to the master station of the DAS to be tested.
- Interaction with the master station of the DAS to be tested and response to the downgoing orders, during which the performance of fault processing is tested.

The master injection testing tool also has the function of importing and exporting the model of the distribution grid from the master station of the DAS to be tested via the data bus following standard IEC-61968. Thus, the modeling of a DAS is strongly recommended to be standard.

In the case where the model is exported from the master station of the DAS to be tested, the testing is aimed at the actual distribution grid to mainly find out the configuration errors. The fault can be set one region at a time automatically so as to test the fault processing performance of every region.

The test may also be based on some specially designed fault scenes, which are imported into the master station of the DAS to be tested. This testing is to verify the adaptability of fault processing performance. It is also used to evaluate DAS products. To tackle lack of fault diversity by setting faults on the actual distribution grids means the test cases should be well designed.

6.4.2 *Secondary Synchronous Injection Testing Methodology and Testing Facilities*

The secondary synchronous injection testing methodology uses the dedicated testing facilities consisting of a commander based on a portable computer and some fault information generating devices.

The secondary synchronous injection testing methodology is illustrated in Fig. 6.7.

A temporary relay protection device should be implanted on the source node in the substation to produce the information of a certain circuit breaker tripped while its corresponding relay protection device acts.

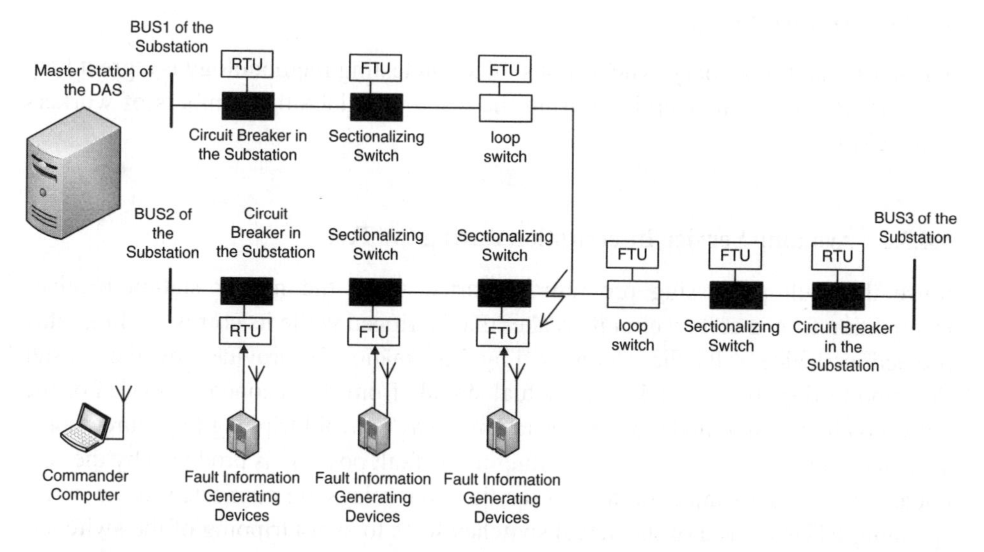

Figure 6.7 Schematic diagram of secondary synchronous injection test method

The fault information generating devices are connected to the terminal units upstream of the fault position to inject the fault current and voltage to the corresponding terminal units simultaneously, causing them to produce fault information to the master station of the DAS to be tested. The fault information generating devices are synchronized by GPS. The testing may cause short-time service interruption of the feeder to be tested.

Before testing, the waveforms of fault currents and voltages are calculated by the analysis software of the commander computer and are sent to the corresponding fault information generating devices.

In the testing procedure, the waveforms of fault currents and voltages are amplified to the secondary ranges of the CT and PT, and injected to the corresponding terminal units simultaneously by the fault information generating devices in the sequence designed beforehand. In other words, the secondary synchronous injection testing facilities produce the simulated fault phenomenon and inject them to the input of the automatic devices.

Thus, the fault processing performance of the coordination of the master station, terminal units, relay protection devices, back-up power supplies, switches, and communication systems can be tested, which is the advantage of the secondary synchronous injection testing methodology. But the workload is quite large during the testing procedure. The numbers of both workers and facilities are great, since each terminal unit needs work and a fault information generating device. Also, the testing may cause short-time service interruption of the feeder to be tested.

6.4.3 Master and Secondary Synchronous Injection Testing Methodology

The master and secondary synchronous injection testing methodology is a great help to avoid of service interruption during the test and reduce the numbers of workers and facilities.

6.4.3.1 Avoiding Service Interruption during the Test

Since the fault processing program is started when the master station receives information from a certain circuit breaker that is tripped while its corresponding relay protection device acts, the corresponding information is provided by the master injection testing tool instead of the actual signals from the secondary devices of the corresponding source node in the substation, so as to avoid tripping the source node. The other fault information corresponding to the fault position is produced by the secondary synchronous injection testing facilities but relays are connected to the corresponding FTUs instead of the actual switches so as to avoid tripping of the switches. Also, the master injection testing tool should be also synchronized by GPS.

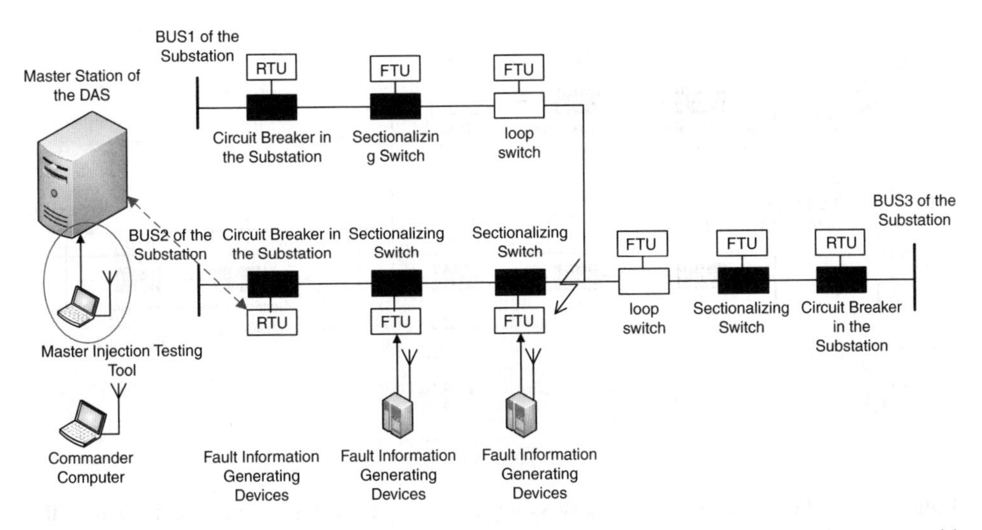

Figure 6.8 The master and secondary synchronous injection testing methodology to avoid service interruption during the test

With this approach, the service interruption is avoided but the coordination of back-up supplies and switches cannot be covered in the test and should be examined by the remote transmission test, which needs a planned outage.

The master and secondary synchronous injection testing methodology to avoid service interruption during the test is shown in Fig. 6.8.

6.4.3.2 Reduce the Numbers of Workers and Facilities

To reduce these numbers, we may connect a few fault information generating devices alternative to the terminal units corresponding to the fault position to inject the fault current and voltage to the corresponding terminal units simultaneously, causing them to produce fault information to the master station of the DAS to be tested. The fault information of the other terminal units corresponding to the fault position is simulated by the master injection testing tool. Thus, only a few fault information generating devices and workers are needed during the test. Of course, the master injection testing tool should be also synchronized by GPS.

The master and secondary synchronous injection testing methodology to reduce the numbers during the test is shown in Fig. 6.9.

In the most compact case, the master and secondary synchronous injection testing methodology only needs one master injection testing tool and one fault information generating device.

With this approach, the numbers of both workers and facilities are reduced but the coordination of backup supplies and switches cannot be covered in the test and should be examined by the remote transmission test, which needs a planned outage.

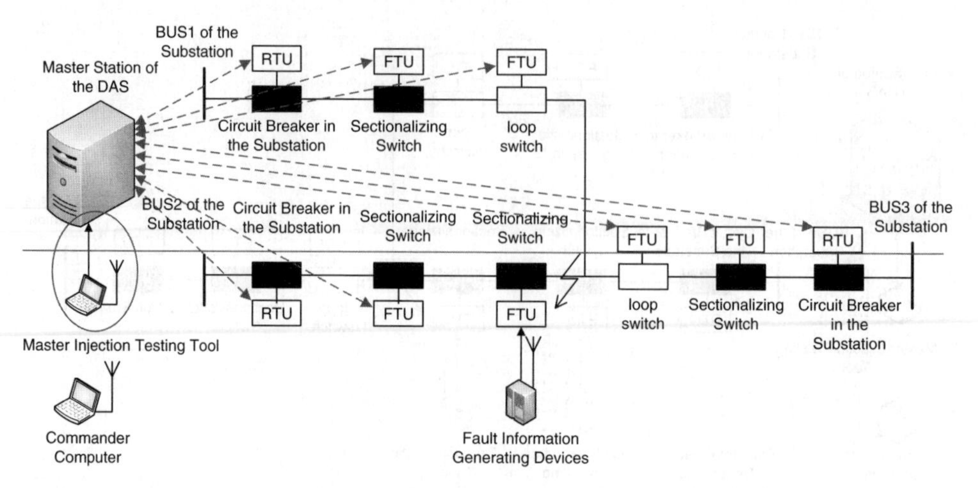

Figure 6.9 The master and secondary synchronous injection testing method to reduce the numbers of workers and facilities

6.4.4 Direct Short-Circuit Test

The three injection testing methodologies described in Sections 6.4.1–6.4.3 are suitable for testing the performance of inter-phase short circuit fault processing for centralized intelligence based DAS and local intelligence devices, but rather difficult to use to test the performance of inter-phase short circuit fault processing for distribution intelligence systems and single phase to ground fault processing. The direct short-circuit test is an effective approach to solve these problems

The direct short-circuit test is to insert a resistor between two phases or one phase to ground of the feeder by a switch to form a real slight fault to test the performance of the fault processing procedure.

The safety protection measures are the key of the direct short-circuit test.

A direct short-circuit test system consists of a specially designed resistor with a high power capacity, a quick switch, a remote controller, a suite of special fittings, and a set of over-current relay protection devices, which is shown in Fig. 6.10.

The resistance of the resistor needs to be carefully designed to meet the requirements of inter-phase short circuit fault and the single phase to ground fault tests. In China, the typical values are 800~1000 Ω for an inter-phase short circuit fault and 100~300 Ω for a single phase to ground fault. As for inter-phase short circuit fault tests, the resistance should be a little larger to reduce electric shock to the system. Thus, sometimes the setting values of corresponding relay protection devices and terminal units may decrease during the test.

The direct short-circuit test is suitable to verify both the inter-phase short circuit fault and the single phase to ground fault processing procedure of any mechanism.

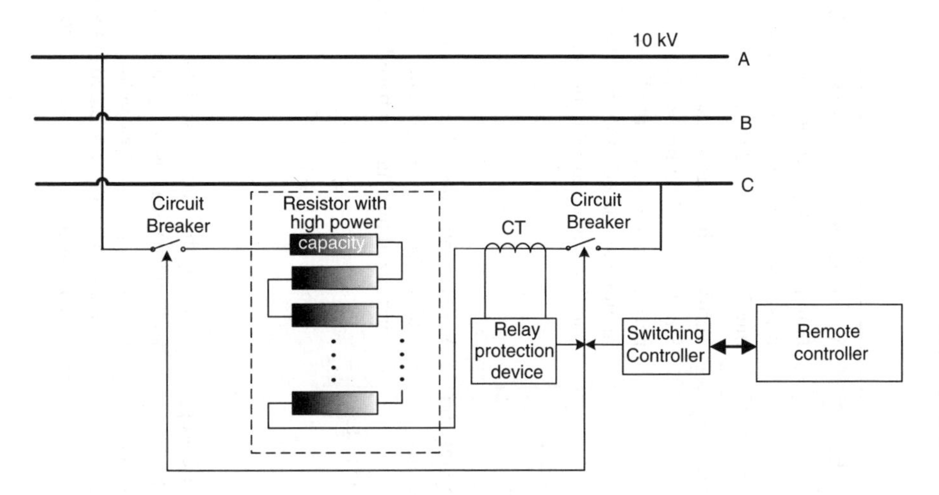

Figure 6.10 A direct short-circuit test system

6.4.5 *Comparison of the Four Testing Methodologies*

The four testing methodologies described in previous sections have their pros and cons. The comparisons are shown in Table 6.5.

The facilities required by the four testing methodologies are shown in Table 6.6.

6.5 Conclusion and Summary

Some approaches including relay protection, automatic reclosing, backup switching, recloser and voltage-delay sectionalizers, reclosing with fast protection, the fast healing approach based on neighbor communication, and centralized intelligence DAS are compared for their inter-phase short circuit fault processing performance. A fault location and restoration approach based on coordination of centralized, distributed, and local intelligence is proposed. When a fault occurs on a distribution grid, the local and distributed intelligent based devices clear the fault and roughly isolate the faulted section immediately. After collecting the total fault information, the DAS accurately locates the fault position and carries out correcting control to finely isolate the fault section and restore the service to as many loads as possible in an optimized way. The performance of fault isolation and restoration for distributed grids can be improved by coordination of centralized, distributed, and local intelligence.

From the point of view of service reliability, the required amounts of various kinds of terminal units in different situations, such as only FTUs with tele-control function, only fault indicators without tele-control function, and hybrids of FTUs and fault indicators, are investigated. The following conclusions are drawn that only one terminal unit for each feeder is the best choice on the view point of cost-benefit ratio, and that the amounts of various kinds of terminal units depend on the desired service reliability

Table 6.5 The comparison of the four testing methodologies

Testing methodologies	Advantages	Disadvantages	Scope of application
Master injection testing methodology	The complicated fault phenomenon and scenes can be set up. Without electric shock. Multiple faults can be simulated for stress test. Without service interruption.	Only covers the master station of the centralized intelligence based DAS. The single phase to ground fault processing performance cannot be tested	Verifying the inter-phase short circuit fault processing performance of the master station of the centralized intelligence based DAS.
Secondary synchronous injection testing methodology	Without electric shock. The coordination of the master station, terminal units, relay protection devices, back-up power supplies, switches, and the communication systems can be tested	The fault phenomena and scenario set up is simple. Multi-fault is difficult to simulate for stress test. The numbers of both workers and facilities are larger. Needs service interruption. The single phase to ground fault processing performance cannot be tested	Verifying the inter-phase Short-circuit fault processing performance of the centralized intelligence based DAS and the local intelligence based relay protections.
Master and secondary synchronous injection testing methodology	The complicated fault phenomenon and scenes can be set up. The master station, terminal units, relay protection devices and the communication systems can be covered. Without electric shock. Multiple faults can be simulated for stress test. Without service interruption. The numbers of both workers and facilities are greatly reduced.	The coordination of back-up supplies and switches cannot be covered in the test. The single phase to ground fault processing performance cannot be tested	Verifying the inter-phase short circuit fault processing performance of the centralized intelligence based DAS.
Direct short-circuit test	More realistic fault phenomena. The coordination of the master station, terminal units, relay protection devices, backup power supplies, switches and the communication systems can be tested. Distributed intelligence based DAS can be tested. The single phase to ground fault processing performance can be tested	The fault phenomenon and scenes set up is simple. Multi-fault is difficult to simulate for stress test. The workload is larger. Needs service interruption and with electric shock.	Verifying the inter-phase short circuit fault and single phase to ground fault processing performance of the centralized intelligence based DAS, the local intelligence based relay protections and the distributed intelligence based DAS

Table 6.6 The facilities required by the four testing methodologies

Testing methodologies	Master injection testing tool	Fault information generating devices	Simulated switches	Direct short-circuit test system
Master injection testing methodology	One	None	None	None
Secondary synchronous injection testing methodology	None	Many	None	None
Master and secondary synchronous injection testing methodology	One	At least one	A few	None
Direct short-circuit test	None	None	None	One

performance, the time to isolate the fault section, the repair time, and the frequency of failure per-year. A DAS planning example is given to illustrate the methodology described in this chapter.

In order to verify the fault processing performance of the centralized intelligence based DAS, the local intelligence based relay protections and distributed intelligence based DAS, four testing methodologies are described including the master injection testing methodology, secondary synchronous injection testing methodology, master and secondary synchronous injection testing methodology, and the direct short-circuit test. The four testing methodologies have their pros and cons and are suitable for use in different applications.

References

[1] Yubo, Y., Yuping, L., Cheng, L., et al. (2006) An adaptive second harmonic restrained method using phase angle and amplitude for inrush detection [J]. *Proceedings of the CSEE*. 26(18):19–24. In Chinese.

[2] Dejun, S., Xianggen, Y., Zhe, Z. et al. (2006) Improved second harmonic restrained method for energizing inrush current [J]. *Power System Technology*. 30(24):84–88. In Chinese.

[3] Kumpulainen, L. K. and K. T. Kauhaniemi (2004) Analysis of the impact of distributed generation on automatic reclosing [C]. *Power Systems Conference and Exposition, IEEE PES*, vol. 1, pp. 603–608.

[4] Jian, L., Xiangqian, T., Xiaoqing, Z. et al. (2014) *Relay Protection and Fault Processing for Distribution Systems*. China Electric Power Press. Beijing. In Chinese.

[5] Jian, L., Zhihua, Z., Xiaoqing, Z. et al. (2011) Relay protection and distribution automation based fault allocation and restoration for distribution systems [J]. *Power System Protection and Control*. 39(16):53–57. In Chinese.

[6] Jian, L., Xinzhou, D., Xingying, C. et al. (2013) *Fault Location and Service Restoration for Distribution Systems*. China Electric Power Press. Beijing. In Chinese.

[7] Jinyi, Y., Jian, L., Li, X. et al. (2014) Distribution grids of high reliability for core areas of large cities. *Automation of Electric Power Systems*. 38(20):74–80. In Chinese.

[8] Zhangqi, W., Nihong, G. (1995) *Switches in Electric Distribution Systems*. China Water and Electric Power Press. Beijing. In Chinese.

[9] Yong, C., Tao, H. (1999) FA based on recloser and voltage-delay type sectionalizers. *Electric Power Technology*. 23(7):31–33. In Chinese.

[10] Jian, L., Rong, D., Wenxue, D. et al. (2002) *Construction and Refination of Distribution Grids for Urban and Rural Areas*. China Water Power Press. Beijing. In Chinese.

[11] Jian, L., Jianli, N. (2003) *New Technologies of Distribution Automation*. China Water Power Press. Beijing. In Chinese.

Fault Location and Service Restoration for Electrical Distribution Systems, First Edition. Jian Liu, Xinzhou Dong, Xingying Chen, Xiangqian Tong, Xiaoqing Zhang and Shiming Xu.

[12] Jian, L., Wei, Z., Hongli, C. (2006) The parameter setting of feeder automation system based on mutual coordination of recloser with voltage–time type of sectionalizers. *Power System Technology.* 30(16):45–49. In Chinese.

[13] Jian, L., Jianzhong, C., Haiyong, G. (2005) A novel feeder automation scheme for rural distribution systems. *Automation of Electric Power Systems*, 29(11):82–86. In Chinese.

[14] Hongli, C., Wei, Z.,Jian, L. (2006) Improvements of recloser and instantaneous protection based feeder automation and its setting approach. *Automation of Electric Power Systems.* 30(15):35–39. In Chinese.

[15] Jian, L., Hongli, C., Qirui, L. (2003) Feeder automation based on recloser and voltage-current mode pole-mounted switches. *Automation of Electric Power Systems.* 27(22):68–71. In Chinese.

[16] Jian, L., Baoji, Y., Qi, C. et al. (2010) A distributed intelligent based feeder automation system with fast self-healing performance. *Automation of Electric Power Systems.* 34(10):82–86. In Chinese.

[17] Jian, L., Shuren, Z., Baoji, Y. et al. (2011) Fast self-healing technology in distributed intelligent feeder automation systems and its reliability enhancement. *Automation of Electric Power Systems.* 35(17):67–71. In Chinese.

[18] Zhenyou, J., Shaohua, J., Wanshun, L. (2002) Theory and analysis of distribution feeder system protection. *Power System Technology.* 26(12):75–78. In Chinese.

[19] Qi, Z. (2002) Disposing fault of distribution system with the elements based on surface protection. *Zhejiang Electric Power.* (2):14–17. In Chinese.

[20] Jian, L., Jianli, N., Yu,D. (1999) A unified matrix algorithm for fault section detection and isolation in distribution systems. *Automation of Electric Power Systems.* 23(1):31–33. In Chinese.

[21] Faguo, Z., Desheng, S., Yu-bin, Y. et al. (2000) Optimized matrix arithmetic of line fault location based on field terminal unit. *Automation of Electric Power Systems.* 24(15):42–44. In Chinese.

[22] Jian, L., Hong-li, C., Hai-Peng, D. et al. (2000) Fault section identification and load balancing of distribution network. *Automation of Electric Power Systems*, 26(22):34–38. In Chinese.

[23] Jian, L., Pengxiang, B., Wenyu, Y. et al. (2007) *Theory and Application for Electric Distribution Grids*. China Water Power Press. Beijing, In Chinese.

[24] Jingdong, L., Changxiu, C., Bangli, Z. (2002) Optimized algorithm of fault section diagnosis for distribution networks based on hierarchical model. *Relay.* 30(8):6–9. In Chinese.

[25] Jun, F., Yuanda, S. 2000, Study of automatic fault location system in the distribution network. *Proceedings of the CSEE.* 20(9):32–34. In Chinese.

[26] Ciric, R. M., Popovic, D. S.(2000) Multi-objective distribution network restoration using heuristic approach and mix integer programming method. *International Journal of Electrical Power & Energy Systems*, Oct. (22):497–505

[27] Baxevanos, I. S., Labridis, D. P. (2007) Implementing multi-agent systems technology for power distribution network control and protection management. *IEEE Transactions on Power Delivery*, 22(1):433–443.

[28] Zhi-Nong, W., Hua, H. E., Yu-Ning, Z. (2002) A refined genetic algorithm for the fault sections location. *Proceedings of the CSEE*, 22(4):127–130. In Chinese.

[29] Si-Qing, S., Zhi-Rui, L., Wen-Qin, Z. et al. (2001) Fault restoration of distribution network based on genetic algorithm. automation of electric power systems, 25(8):53–55. In Chinese.

[30] Ya-Ming, S., Zhi-Wei, L. (2003) Assessment of data mining model based on the different combination rough set with neural network for fault section diagnosis of distribution networks. *Automation of Electric Power Systems*, 27(6):31–35. In Chinese.

[31] Trovato, M., Delvechio, G., Bualoti, R. (1999) A fuzzy reasoning approach for distribution automation. *Power Technology, Budapest 1999, IEEE*, (1):249.

[32] Hsu, Y.-Y., Kuo, H.-C. (1994) A heuristic based fuzzy reasoning approach for distribution system service restoration. *IEEE Trans. On Power Delivery*, 9(2):948–953.

[33] Ying-Ying, W., Yi, L., Guang-Yu, T. (2005) Fault location based on Bayes probability likelihood ratio for distribution networks. *Automation of Electric Power Systems*, 29(19):54–57. In Chinese.

[34] Jian-Xin, C., Jian, L. (2003) An uncertainty reasoning system based on trouble call information for location faults on distribution network. *Proceedings of the CSEE*, 23(4):57–61. In Chinese.

[35] Jian, L., Qian, Z., Hongli, C. et al. (2010) Robust fault diagnosis and restoration for distribution grids. *Automation of Electric Power Systems*. 34(7):50–56. In Chinese.

[36] Jian, L., Xinzhou, D., Xining, C. et al. (2012) Robust fault isolation and restoration for distribution systems. *Power System Technology*, 36(1):253–257. In Chinese.

[37] Zhaoqiang, G., Guoqing, T., Lei, W. (2000) Integrated intelligent service restoration system for distribution network– An auto-learning fuzzy expert system combined with service restoration algorithm set. *Automation of Electric Power Systems*, 24(2):17–21. In Chinese.

[38] Jingliao, S., Yongli, L., Shengwei, L. et al. (2009) A protection scheme for distribution system with distributed generations. *Automation of Electric Power Systems*, 33(1):81–84. In Chinese.

[39] Wei, H., Jinyong, L., Xiang, X. et al. (2008) Influence of distributed generation on phase-to-phase short circuit protection in distribution network. *Automation of Electric Power Systems*, 32(1):93–97. In Chinese

[40] Morren, J., de Haan, S. W. H. (2007) Short-circuit current of wind turbines with doubly fed induction generator. *IEEE Transactions on Energy Conversion*, 22(1):174–180.

[41] Aoyang, H., Zhe, Z., Xianggen, Y. (2009) The simulation research on the characteristic of fault current of doubly-fed induction generator. *Journal of Huazhong University of Science and Technology*, 37(9):105–108. In Chinese.

[42] Jinxin, O., Xiaofu, X. (2010) Characteristics of short-circuit current of wind turbine driven DFIG and its impacts on grid-connected distribution network. *Automation of Electric Power Systems*, 34(23):106–110. In Chinese.

[43] Jian, L., Shuren, Z., Xiaoqing, Z. et al. (2013) Fault allocation for distribution systems with distributed generations. *Automation of Electric Power Systems*, 37(2):36–42, 48. In Chinese.

[44] Jian, L., Tao, L., Long, L., et al. (2014) Adaptability of distribution automation systems to photovoltaic installation. *Protection and Control of Electric Power Systems*, 42(20):7–12. In Chinese.

[45] Ciric, R. M., Popovic, D. S. (2000) Multi-objective distribution network restoration using heuristic approach and mix integer programming method. *International Journal of Electrical Power & Energy Systems*,Oct. 2000(22):497–505.

[46] Perez-Guerrero, R., Heydt, G. T., Jack, N. J., et al. (2008) Optimal restoration of distribution systems using dynamic programming. *IEEE Transactions on Power Delivery*, 23(3):1589–1596.

[47] Miu, K. N., Chiang, H.-D., Yuan, B. (1998) Fast Service restoration for large-scale distribution system with priority customers and constraint. *IEEE Transactions on Power Systems*, 13(3):789–795.

[48] Kleinberg, M. R., Miu, K., Chiang, H. D. (2011) Improving service restoration of power distribution systems through load curtailment of in-service customers. *IEEE Transactions on Power Systems*, 26(3):1110–1117.

[49] Jian, L., Xiaoqing, Z., Zhihua, Z. et al. (2013) Location and restoration of the short-circuit with two phase grounded to the earth in distribution systems. *Automation of Electric Power Systems*, 37(5):105–110. In Chinese.

[50] Chiang, H.-D., Nan Miu, K. (1996) Parallel genetic algorithm for service restoration in electric power distribution systems.*Electrical Power & Energy Systems*, 18(2):111–119.

[51] Kumar, V., Kumar, H. C. R., Gupta, I., et al., (2010) DG Integrated approach for service restoration under cold load pickup., *IEEE Transactions on Power Delivery*, 25(1):398–406.

[52] Santos, A. C., Delbem, A. C. B., London, J. B. A., et al., (2010) Node-depth encoding and multi-objective evolutionary algorithm applied to large-scale distribution system reconfiguration. *IEEE Transactions on Power Systems*, 25(3):1254–1265.

[53] Augugliaro, A., Dusonchet, L., Sansevvrino, E. R. (2000) Multi-objective service restoration in distribution networks using an evolutionary approach and fuzzy sets. *Electric Power and Energy Systems*. 2000(22):103–110.

[54] Kumar, Y., Das, B., Sharma, J. (2008) Multiobjective, multiconstraint service restoration of electric power distribution system with priority customer. *IEEE Transactions on Power Delivery*, 23(1):261–270.

[55] Chen, C.-S., Lin, C.-H. (2002) A rule-based expert system with colored Petri net models for distribution system service restoration. *IEEE Transactions on Power Systems*, 17(4):1073–1080.

[56] Lin, W. M., Chin, H. C. (1997) Preventive and corrective switching for feeder contingencies in distribution system with fuzzy set algorithm. *IEEE Trans on Power Delivery*, 12(4):1711–1716.

[57] Zhigang, L.,Yuxiang, D. (2006) Distribution system restoration based on improved binary particle swarm optimization. *Automation of Electric Power System*, 30(24): 39–43. In Chinese.

[58] Jian, L., Shuren, Z., Xiaoqing, Z. et al. (2011) Critical techniques of fault isolation and restoration for distribution systems. *Automation of Electric Power Systems*, 35(24):74–79. In Chinese.

[59] Jian, L., Zhihua, Z., Xiaoqing, Z. et al. (2011) Research of modeled fault isolation and restoration for distribution systems. *Power System Technology*, 35(11):97–102. In Chinese.

[60] Jian, L., Jingqiu, X., Hongli, C. (2004) Algorithms on fast restoration of large area breakdown of distribution systems under emergency states. *Proceedings of the CSEE*, 24(12):132–138. In Chinese.

[61] Genjun, C., Li, K. K., Guoqing, T. (2002) A tabu search approach to distribution network reconfiguration for loss reduction. *Proceedings of the CSEE*, 22(10):28–33. In Chinese.

[62] Jian, L., Xiao-Jun, S., Hong-li, C. et al. (2008) Restoration for large area blackout of distribution network and switching operation sequence management. *Automation of Electric Power Systems*, 32(2):77–83. In Chinese.

[63] Dong, L., Yunping, C., Guang, S. et al. (2006) CSP-based model and algorithm of service restoration for large area blackout of distribution system. *Automation of Electric Power System*, 30(10):28–32. In Chinese.

[64] Yu-qin, X., Li, Z., Zeng-Oing, W. (2009) Agent-environment-rules model based algorithm of service restoration for large-area blackout of distribution system. *Power System Technology*, 33(12):66–71. In Chinese.

[65] Jeerings, D. I., Lindersa, J. R. (1991) Protective relay for down-conductor faults. *IEEE Transaction on Power Delivery*, 62(2):565–574, April.

[66] PSERC Working Group D15, "High Impedance Fault Detection Technology", March 1, 1996.

[67] Aucoin, B. M., Jones, R. H. (1996) High impedance fault detection implementation issues. *IEEE Transactions on Power Delivery*, 11:139–148.

[68] Yushan, H., Yihan, Y., Yuanheng, R. (1994) Micro-Processor based single-phase-to-ground fault line selection based on amplitude and polarity comparison in the system with neutral non-effectively grounded. *Information on Electric Power*, 2:15–19.

[69] Dong, X., Shi, S. (2008) Identifying single-phase-to-ground fault feeder in neutral non-effectively grounded distribution system using wavelet transform,*IEEE Transactions on Power Delivery*, 23(4):1829–1837. Oct.

[70] Dong, X., Bi, J. (2005) Analysis of transient travelling wave and study on fault line selection for distribution lines, *Proceedings of the CSEE*, 4:1–6.

[71] Bi, J., Dong, X., Zhou, S. (2005) Fault line selection based on two-phase current traveling wave, *Automation of Electric Power Systems*, 50(3):17–20.

[72] Shi, S., Dong, X. Traveling waves based on-line insulation monitoring of power cable, *National Invention Patent of China*, ZL200910081348.7.

[73] Shi, S., Dong, X., Traveling waves based fault forecast in power system, *National Invention Patent of China*, ZL 200910081349.1.

[74] Shi, S., Dong, X. Single-phase-to-ground fault location method and its device, *National Invention Patent of China*, ZL 201010244606.1.

[75] Shi, S., Dong, X., Zhou, S. (2005) Analysis of single-phase-to-ground fault generated travelling waves, *Automation of Electric Power Systems*, 53(23):29–32.

Index

Fault Location and Service Restoration for Electrical Distribution Systems, First Edition. Jian Liu, Xinzhou Dong, Xingying Chen, Xiangqian Tong, Xiaoqing Zhang and Shiming Xu.